"Nikolay Anguelov's new book makes an important contribution to fashion sustainability scholarship through a critical analysis of the cultural legacy of the fashion industry. It focuses on the historical developments that have created the current industrial reality of unsustainable practices to provide an argument for the need for major cultural changes in consumption and commerce. The book enhances our understanding of what kind of change can support innovation aspects in business and policy towards sustainability."

Kirsi Niinimäki
Associate Professor of Design, School of Arts, Design and Architecture, Aalto University, Finland

"Anguelov continues to be a leading voice on sustainable fashion and a highly articulate critic of the industry's existing environmental platforms. In remarkable detail and with methodological rigor, he meticulously outlines global apparel supply chains from cotton fields to the fashion shops and retail outlets of the global economy. The picture that emerges is one in which we all bear some responsibility. This includes the young consumers intent on wearing the latest fashion trend promoted by their favorite influencer to the government regulators incapable or unwilling to control the global industry.

"Anguelov saves his most biting critique for fast fashion executives who have built up this industry by aggressively promoting mass consumption and made billions by failing to pay the 'social costs' of public health damage, loss of bio diversity, and climate change. He calls out unacceptable claims by fashion brands for what they are: greenwashing. This includes H&M's assertion that, by 2040, it will be climate positive by capturing more CO2 emissions than its core supply chain emits, a claim he calls 'preposterous.' Anguelov ends on a note of hope that concerned citizens, fashion consumers, and advocacy groups can use their political voice for policy change. This book is highly recommended for everyone concerned about our planet and interested in addressing the fast fashion sustainability crisis."

Mark Anner
Professor of Labor and Employment Relations, and Political Science
Director, Center for Global Workers' Rights
Director, MPS Program in Labor and Global Workers' Rights
(part of the Global Labour University network)
The Pennsylvania State University

"This is a timely and valuable contribution to the burgeoning literature on fashion (un)sustainability. The author's 2015 book was at the vanguard of the sustainable fashion movement as one of the first to shed light on the social and environmental cost of fast fashion, and this one is likewise a much-needed addition that makes sense of the wealth of information and publications that now exist. In recent years we have witnessed a surge in interest from multiple perspectives including the media, academia and grassroots organisations, to name a few, resulting in a wealth of information, but also a fair amount of greenwash in one form or another. Sustainable fashion has become a buzzword. Fashion brands and retailers are increasingly sharing information about production and supply chain management, but meanwhile expanding their operations globally, speeding up the frequency of new collections and producing ever greater volumes of stock. Furthermore, a more nefarious form of fast fashion has emerged from a new breed of online-only ultra-fast fashion retailers who make heavy use of sophisticated digital marketing tactics to promote an ever-changing array of trendy items at pocket-money prices. As such, there is no topic on which a treatise is more needed to take stock of the current state of play and a look forward towards viable interventions in business, science and policy that could support a transition to a sustainable industry.

"This is an authoritative, detailed and evidenced work which takes a deep and critical dive into the paradoxes and complexities of (un)sustainable fashion. Written in an engaging style with numerous original perspectives, it is ideally suited to fashion students in a multitude of disciplines from design to marketing to textile engineering, as well as those working in the industry across various functions and members of the general public who have an interest in the topic. The book clearly sets out the complex realities of the global fashion industry and explains the cultural, economic and political reasons for its social and environmental impacts in the key areas of waste, carbon footprint, pollution and exploitation. The arguments are critical, balanced and applied using evidence from research across a range of relevant disciplines such as chemistry, material and environmental sciences, public policy, engineering, data management, marketing and finance – clearly showing that fashion is bigger than you may at first think and that sustainable solutions require multiple stakeholder inputs. For those who are interested to find out more, there is a wealth of references to both classic and recent important scholarship. The book concludes with a refreshingly critical analysis of specific interventions that could support a transition towards sustainability, such as business model innovation, circularity and forms of governance, including an honest consideration of the impact of COVID-19 on the willingness and ability of the sector to transform."

Patsy Perry
Professor of Fashion Marketing at Manchester
Fashion Institute, Manchester Metropolitan University

"Nikolay Anguelov's *The Sustainable Fashion Quest* is a rich account of how we have arrived at today's crisis of fast fashion. Anguelov cuts through the greenwashing of well-known fashion brands, and by grounding the book in current research he lays out the issues of overproduction in a global context. An overview of legislative efforts and various self-regulation initiatives to tackle the crisis completes the book. Reading it invites the difficult, vital questions: how do we transition from a world in which insatiable greed drives the fashion business to one that puts the Earth and her inhabitants, including all of humanity, first?"

Dr. Timo Rissanen
Associate Professor of Fashion and Textiles,
University of Technology Sydney

The Sustainable Fashion Quest

The Sustainable Fashion Quest

Innovations in Business and Policy

Nikolay Anguelov, PhD

A PRODUCTIVITY PRESS BOOK

First Edition published 2021
by Routledge
600 Broken Sound Parkway #300, Boca Raton FL, 33487

and by Routledge
2 Park Square, Milton Park, Abingdon, Oxon, OX14 4RN

Routledge is an imprint of the Taylor & Francis Group, an informa business

Library of Congress Cataloging-in-Publication Data
A catalog record for this title has been requested

ISBN: 978-0-367-72076-6 (hbk)
ISBN: 978-0-367-72073-5 (pbk)
ISBN: 978-1-003-15334-4 (ebk)

Typeset in Garamond
by SPi Global, India

Contents

Prologue

This book is a follow-up to Anguelov, N. (2015): *The Dirty Side of the Garment Industry: Fast Fashion and Its Negative Impact on Environment and Society*. When writing this prequel, a process which started in 2008, I felt like a lone voice. Not a lot was written about fashion sustainability, much less from a critical perspective of the social embrace of the commerce of cheap clothes. Much has changed. And yet, not really.

It appears that everyone is talking about sustainability in fashion. In the past five years books, initiatives, innovations, and even governance structures have emerged, urging apparel brands to decarbonize. The term itself, "decarbonize," I came across last year. I started the research process with a hopeful motivation to provide an overview of this positive social movement toward changing the industry. I had to learn a lot as a staggering amount of academic research has emerged in such a short time. From works on sustainable fashion retail practices to chemical research on cleaner production in fabric manufacturing to political and legal works on new laws for the making and disposal of clothes, it is a body of work that basically did not exist when I was doing research for *The Dirty Side of the Garment Industry*.

Covering the findings of the literature was a massive undertaking, as the volume is immense and comes from across academic disciplines. The list of references here alone is the length of a short academic book. As an advocate, this fact fills me with hope. Social awareness of a problem is the pre-requisite for a cultural change. It is needed to support transformations in business and policy to address the problem, in this case it being the high carbon and toxicity footprint of the fashion industry.

During the research process, I relied heavily on input from industry professionals to keep me abreast on implementations of innovations in production and operations management. My main source of insight was Ariel Kraten, Director and co-founder of *GoBlu* – a sustainability accelerator service provider to textile and apparel companies. Since 2017, Ariel and I have collaborated in presentations, university course development, and social outreach initiatives. I humbly try to put into this volume my learning from her professional experiences as a practitioner on how the industry is changing. Ariel has guided the choice of topics and material covered in this book, helped with clarifications on technical and chemical processes and offered first-hand examples from her interactions with textile factories and brand managers from around the world.

Our hope for this book is to explain the changes that are happening in the fashion industry. Much is changing for the better, yet much more change is needed. Therefore, it is with great pleasure that we note the emergence of numerous academic programs in universities around the world dedicated to fashion sustainability. The students in those programs will be essential in implementing viable changes in the future. We are grateful they are enrolling in such programs and I hope the information in this book on fashion culture, dynamics, economics, policies, and politics will help them on their way to transforming the industry.

Authors Bio

Nikolay Anguelov, PhD is Associate Professor in the Department of Public Policy at the University of Massachusetts, Dartmouth. He holds a doctorate in Policy Studies, concentration Regional Economic Development, a Master of Public Administration (MPA), as well as a Master of Science in Applied Economics and Statistics from Clemson University. Dr. Anguelov completed his undergraduate studies in International Trade and Marketing at the Fashion Institute of Technology, which provided the foundation of his knowledge of fashion economics. As a political economist, he focuses on the intersection of economics and politics, both in the US and internationally. His research links economic development, globalization, sustainability, and public finance. He is the author of *From Criminalizing to Decriminalizing Marijuana: The Politics of Social Control* (2018, Lexington Books), *The Dirty Side of the Garment Industry: Fast Fashion and Its Negative Impact on Environment and Society* (2015, CRC Press), *Economic Sanctions vs. Soft Power: Lessons from North Korea, Myanmar and the Middle East* (2015, Palgrave Macmillan), and *Policy and Political Theory in Trade Practice: Multinational Corporations and Global Governments* (2014, Palgrave Macmillan).

Ariel Kraten is co-founder and director of GoBlu International, an industry-leading sustainability service provider for textiles and apparel manufacturers. She works with dozens of global brands to deliver their corporate sustainability strategies with a focus on internal buyer/designer education and communication. Miss Kraten oversees a unique training program for factories in major sourcing regions including Bangladesh, China, and Ethiopia. She has supported global multi-stakeholder initiatives in developing and implementing tools for reducing the environmental impact of the apparel industry. Also a lecturer at the Experimental College of Tufts University, Ariel takes inspiration and optimism from the students in her course #Outfitoftheday: Clothing, Sustainability, and the Global Implications of Getting Dressed.

Introduction

A Letter to the Fashion Student

Congratulations! You are on the road toward an exciting and gratifying career in textiles and fashion. Regardless of where your particular interests lie, we welcome your passion for this unique and complex industry. Students may be inspired to study fashion because of the perceived glamour, luxury, whimsy, creativity, and splendor of the industry. While these factors do have a role to play, working in fashion is much more complicated than picking out outfits, creating style combinations, or producing glamorous advertising campaigns. The reality is that fashion is a serious global industry with significant social and environmental ramifications. While fashion products are design-driven, and creativity is certainly prized, the broader industry relies heavily on skill sets and sectors that students may not expect, such as chemistry, material and environmental sciences, public policy, engineering, information technology, marketing, and finance.

In truth, we in the industry need your talents in all of these areas and more. Beyond your intellect and skills, we need your humanity and your leadership. At this pivotal moment in the evolution of the fashion industry, the most valuable people will be those who are willing and able to critically examine their own roles, their company's policies and structural status quos to push back against practices that lead to ecological and social injustice. This is how we will change the trajectory of the fashion industry, which is notorious for its impact in terms of waste, greenhouse gas emissions, water pollution, ecological destruction, biodiversity loss, and unsafe, exploitative, and sometimes fatal working conditions.

As a student, and then as a young professional, you may question your ability to impact these looming global issues. However, a truth unique to this industry is that individuals in early-career positions have both power and leverage. Young buyers may find themselves holding the purse strings and making the call as to whether to work with the supplier with the outstanding environmental scorecard or the one with the quicker sample turnaround time. That is a powerful seat from which to affect change.

To be able to do so, you need to understand the challenges the industry is up against. That is our goal with this text. We hope to equip you with the information you need to see the ways your company may be contributing to the problematic issues of current fashion economics, empower you to question such activities, and facilitate in finding solutions.

Consider this hypothetical example: A brand decides to pay its partner factory, which is in charge of dyeing fabric, in four installments rather than in one upfront payment, which would cover the costs of purchasing the required dyestuff and material. This means that the factory is not able to buy and process everything at once, and dyeing needs to be done in four separate batches and between other runs for other clients. However, small changes in weather, auxiliary chemicals, or a changeover in "dye master" – the person in the factory in charge of mixing dyes – tend to create small differences in the resulting color. The buying office receives the fabric, notices the color variability, and then demands that the dyehouse re-dye some of the materials.

There are added costs to this process that put extra strain on the dyeing mill. Additional chemicals must be procured, or perhaps even fresh material needs to be purchased and re-dyed from scratch. Workers need to duplicate their efforts, including establishing all the proper paper trails the brand requires regarding chemical use and raw material origins. Furthermore, the unexpected time needed for the re-dye means that employees may be required to work extra hours so that other tasks for other clients do not fall too far behind.

It is easy to see how a cascade of negative effects can result, and how these effects could impact working hours, pay, and even safety. However, buyers within a brand's sourcing office may have never considered the complications that may result from paying their suppliers in installments rather than upfront and in full. They may just be following a company policy regarding payment terms and such company policies are often set by financial management consulting firms, helping brands with their balance sheets. The financial decision makers who would advise the company's chief financial officer on how to maximize the firm's liquidity for its quarterly profitability evaluations, aimed to impress investors and shareholders, would know finance, not fabric dyeing.

Maybe you, as a future buyer, would know about the lifeblood of the industry, which is fabric. Your job would be to know it the way a baker knows dough – all its inputs, its properties, its functions, and all that is needed for your clients to manufacture it. Your job would also be to communicate across your company's structure about what is needed for your suppliers to produce it efficiently, expediently, and sustainably.

The industry is vowing to dramatically change how fabric is manufactured. It is at a crossroad of transformation. Additionally, the structural weaknesses highlighted by the coronavirus have wreaked havoc on so many people's livelihoods, but this upheaval has also created a mandate for change.

We must seize this momentum. It will take a commitment from people like you to evolve the business away from focusing only on the bottom line to owning a responsibility to humanity and the planet. We need your innovative and entrepreneurial thinking in creating a new set of incentives in fashion – for consumers, for corporations, and, crucially, at the policy level. We hope that this text will equip you with the information you will need to become a change agent from within.

Those of us already dedicated to this challenge welcome you to the team, whether you bring an eye for design, a specialty in engineering, expertise in chemistry, or innovative ideas that can transform the business model itself. Use your position, wherever that may be, to help bring about a socially responsible and environmentally respectful industry, where the goods you help create not only bring joy to consumers, but sustainable opportunities for those workers the industry relies upon and a nurtured planet for us all.

Warmly,
Nikolay and Ariel

Chapter 1

The Evolution of Fashion Business Models

From Value Retail to Fast Fashion

Today, the fast fashion business model has come to define the industry. The largest brands in the world, in terms of market share and monetary value, are the leading fast fashion conglomerates *H&M* and *Zara* (*Inditex*). These companies lead in creating a fashion economy based on the reliance of retailing very affordable items. The selling of cheap clothes is the main profit stream in fashion economics.

Fast fashion retail has grown from a specific form of merchandising by a few brands to a necessary product line category for all brands. As a retail tactic, fast fashion was pioneered by *H&M* and revolutionized by *Zara* over a decade ago (Mo, 2015; Tokatli, 2008). It transformed fashion label culture, stripping brands of long-established statuses, and cultural meaning. It changed how consumers perceive the very essence of the value of fashion purchases. That value had traditionally been most directly linked to prestige.

Fast fashion erased the prestige hierarchy of the industry, which had been slowly and carefully built for a century to define the very core meaning of the term "fashion" as signifying social status – a status of prestige. Fashionable items bring to their owners an emblem of excellence: excellence of artistic value, as signified by the design of the items; excellence in quality of product, as showcased by the materials used; and an excellence of social status. It is the redefinition of this excellence of social status that is revolutionizing the industry today.

Historically, showcasing excellence of social status was a function of the fact that fashionable items were expensive. Owning them was a status symbol, signifying a belonging to a higher social class. This prestige dynamic was erased by the advent and proliferation of the fast fashion business model. Fast fashion altered prestige dynamics through the core attribute the fashion industry had always used to demarcate its iconographic relevance in defining social status – the price of clothes.

Since its onset, the fashion industry has had an incentive to retail product at high prices. Although clothes are an everyday necessity, the history of how they are made, sold, and worn indicates that in terms of affordability, clothes were not cheap at the onset of fashion industry. Whitaker (2006) offers a chronological history of the price evolution of fashion retail with examples from the earliest product positioning in department stores. The history captures the evolution of retail in America and Western Europe with explanations that even the most basic apparel items – lingerie to ready-made dresses and basic business suits – had ranged in prices comparable to today's equivalencies from $70 for lingerie to $169 for women's suit combination. Department store culture – also referred to today as "traditional retail" – had an incentive to keep prices as high as possible. Therefore, it is known that for the majority of the industrial age's history of clothes, the most worn pieces of apparel were hand-me-downs (Rosenthal, 2007). In addition, until the 1980s, even mass-market prices were kept at such high levels as to create incentives for women to sew their own clothes, mend old clothes, and generally keep clothes for as long as possible (Buckley, 1998).

This history reflects the retail of mass-market apparel. High fashion has operated at price levels of exclusivity for centuries and continues to do so. Yet, today high fashion is not the industry's prestige emblem. Affordability is. The fast fashion revolution elevated price competitiveness to the forefront of important attributes in the apparel trade.

The change agents – the first fast fashion global conglomerates – *H&M*, *Zara*, *Forever 21*, *Top Shop* – today are more than just well-known brand names. They enjoy unequal and unprecedented global brand awareness as trendsetters, style dictators, and fashion authorities. This status is unprecedented because since the beginning of "value retail" – the commerce of apparel at price points below those in traditional department stores – value retailers sold cheaper clothes without any claim as to their being fashion-forward (Pettinger, 2004). Value retailers followed the trends put forth by high fashion.

Value retail evolved alongside the "branded retail" model – best exemplified by the traditional department store – as a low-price alternative for selling merchandise to, mainly, younger buyers (Lea Wickett, Gaskill, & Damhorst, 1999). Brands such as the *GAP, American Eagle, Abercrombie & Fitch, The Limited*, and *J. Crew* grew in popularity during the era of branded retail by changing the industry's focus from the traditional market demographic – female buyers aged 21 and over (Crane, 1999; Easey, 2009; Parment, 2013) – to younger market segments. Cook and Kaiser (2004) offer a thorough historical evolution of the refocusing of the industry toward an increasingly younger demographics in the context of teen sexualization. Notedly socially indefensible, the commercialization of this trend emerged as a retail response to cultural and style changes in the 1960s and 1970s, with the proliferation of such items as the t-shirt and the blue jeans. For example, *ESPRIT* was founded in 1968 and the *GAP* in 1969, both in San Francisco, CA.

The new value retailers emerged to sell increasing numbers of t-shirts and blue jeans to buyers younger than 21. What the industry realized, which enabled both the redirection toward younger buyers and the rise of value retail, is that these younger consumers may have lacked the purchasing power and disposable income of the older buyers, but not their commitment to dress stylishly and fashionably. Technology was also evolving rapidly, bringing about mass media, mass-entertainment, and mass-culture. What was stylish and fashionable was promoted via increasing variety and scope of media platforms, reaching wide audiences. A new professional workforce emerged of market analysts, trend prediction agencies, and industrial psychologists, to systemically evaluate the consumer behavior of fashion buyers.

With all the technological innovation, social emancipation, and business segment augmentation, during the era of traditional apparel retail, branded and value retailers remained fashion trend followers. They merely commercialized the trends put forth by the few fashion houses that had earned their status as cultural fashion dictators through decades of building their labels' status, based on exclusivity. The fashion labels, of course, had help from the contingent industrial sectors that profited from fashion promotion – trend forecasting agencies, the editors of fashion magazines, and department stores (Crane, 1999). The tandem of this collusion set forth the trends that the lower tier retailers brought to the mass market and all its different demographic segments. The main economic power of the industry – the mass market – was told what was fashionable and what was not, when, and for how long.

The mass-market retailers may not have been the innovators of fashion style, but they were the innovators of selling it. Customer service, in and out of store advertising, and multi-media promotion to sell a shopping experience, not just product, gradually became knowledge streams in retail competition. Promoting the shopping experience became central to building retailer brand awareness. The appropriate term emerged to describe this business model – branded retail (Carpenter & Fairhurst, 2005). Retail innovation in fashion evolved as a function of identifying opportunities to increase sales in different market segments by understanding their consumer behavior and response to this culture of top-down fashion dictatorship. It was such innovations that brought about fast fashion commerce.

Fast fashion was a product of two decades of retail learning. The impetus for the morphing of value retail to fast fashion was competition to increase sales as retailers themselves grew. Department stores became chains and branded and value retailers were "born global," as is the term, to describe their business model of not just fragmented production but also retail strategy of selling around the globe in hundreds, if not thousands, of stores (Bair & Gereffi, 2004; Bell, McNaughton, & Young, 2001).

In this global market place, the competition evolution escalated and continues to evolve on two fronts – product differentiation and price differentiation (Beath & Katsoulacos, 1991; McColl & Moore, 2011). Ever increasing varieties of product options are sold not just as luxury or mass-market goods, but as differentiated product lines positioned at strategic "price point" levels for consumers with different willingness to spend. Between luxury and mass-market items emerged the retail of "prestige" product lines and "masstige" product lines. Product differentiation even ventured below the cutoff level of low prices in mass-market retail – cheaper than mass-market options emerged, politely referred to as "budget" product lines, not to be confused with "value" product lines. The difference is that "budget" product can be and is often retailed inside a "value" store. Who doesn't love the sales section at the *GAP*? And what would *GAP* sales be without building in its customers the knowledge that although they walk in the store expecting low prices, they can find even lower prices than they might expect? This type of retail, referred to as the "in-store surprise" model, gradually became the cornerstone of both the branded and value retailers (Lund, 2015; Muruganantham & Bhakat, 2013; Pomodoro, 2013).

Since the 1970s, outside of luxury retail, the gradual trend was to create a culture of buying based on an impulse as a function of surprise (Joo Park, Young Kim, & Cardona Forney, 2006; Pentecost & Andrews, 2010). The surprise was not the product itself, but its price. It became a mantra in retail management to posit that in order to be successful, retailers had to use a strategy of overstimulation to engender unplanned purchases in their consumers (Han et al., 1991). The tactics employed became numerous in the traditional physical store – from club music and mood lighting, to semi-clad salespeople in *Abercrombie & Fitch*, to the "sales section" at the *GAP* that somehow had merchandise that seemed anything but cheap. There is indeed academic research from fashion advertising on the effect of in-store lighting on brand allegiance and image (Schielke & Leudesdorff, 2015).

It quickly became apparent to retailers that price was more attractive to consumers than style. And yet, the fashion houses still managed to hold on to their leadership as trendsetters, expanding globally by opening boutique stores in exclusive locales around the world and moving into differentiated product development but not of clothes. They maintained their global brand awareness through selling cheaper product options for their expensive clothing lines, which were the accessories, cosmetics, and fragrances that could be branded but retailed at price points much lower than luxury apparel. After all, there are only so many thousand-dollar dresses a retailer can sell each season. There are many more hundred-dollar bottles of perfume and twenty-dollar kits of make-up that status shoppers are willing to buy.

As high fashion label commerce became global and based in investor finance, the need to increase marginal sales in order to meet investor return-on-investment (ROI) quarterly expectations, became a core goal of profitability. To meet it, high fashion labels increased the type of non-apparel product lines they sold, particularly cosmetics and fragrances (Nueno & Quelch, 1998; Priest, 2005). Tungate (2008) dedicates chapter 13 of *Fashion Brands: Branding Style from Armani to Zara* to the dynamic, with the appropriate subheading on page 157: "Brand in a Bottle." The fashion industry became renamed: "the fashion and related industries."

The fashion labels went global and their designer superstars became cultural icons as global brands themselves. Globalization was bringing growing numbers of customers from the developing world into the industry's profit range. Those customers, looking for links to an identity as global citizens (Anholt, 2006; Steenkamp, 2014; Van Gelder, 2005), helped high-fashion brands stay on the forefront of style leadership. It was the strength of their brands, in terms of global brand awareness, that international audiences favored. It allowed buyers to feel connected to cultural legacies outside of their respective countries of origin. It defined the global shopper as one of international cultural identity.

Purchasing products with global brand status became the standard buying behavior for many consumers. The term "cultural convergence" emerged to define this confluence of cultural consumption (Leung et al., 2005; Lynch, & Strauss, 2007; Snyder, Willenborg & Watt, 1991). The problem in the "fashion" industry was that the consumption was of mainly non-apparel items that carried the coveted brand labels – the bottles of perfume, cases of make-up, and sticks of lipstick and mascara. Hence, the joke emerged in professional circles on the difference between working in fashion and working in apparel. It had to do with what was your job to sell. If your job was to sell clothes, you worked in the apparel industry. If you worked in "fashion," your job was to sell anything but clothes. In either, it was your job to sell "what customers don't know they want yet" – an often-repeated phrase in the industry now, first uttered by Diana Vreeland, legendary *Vogue* fashion editor of the 1960s and 1970s (Vreeland, Tcheng, & Perlmutt, 2011).

It is thus easy to see the main factor that led to the dethronement of the high-end fashion and branded apparel labels as the global style trendsetters by the fast fashion retailers. They were not focused on selling clothes. The fashion labels dictated down to mass-market consumers the "styles" and "fashionable" looks through integrated images in visual promotions of emotions and personalities. However, they did not directly design clothes to sell to the average consumer. Fast fashion started with meeting this core market need – provide affordable, yet fashionable, clothes with the mass-market consumer in mind – and at each of the spiraling-down price points of the different product line categories. The one thing fast fashion retail needed to do is establish its credibility in style leadership. The way it achieved that goal was by refocusing the industry's attention and understanding of "fashionable" toward novelty.

From Exclusivity to Novelty: Redefining Style Innovation

The novelty of style had always been the core feature in the competitive advantage of successful fashion design. That is why it was so carefully guarded by the three traditional fashion brand types of corporation – the "fashion house," the "luxury designer brand," and the "high-fashion" active-wear brand. The fashion houses, such as *Louis Vuitton, Dior, Chanel, Givenchy, Gucci, Prada, Fendi*, had been in existence since the late 19th and early 20th century, remaining in the forefront of trendsetting by hiring the best designers at the time. Each aspiring designer entered the world of design hoping one day to rise to the ranks of chief designer for these palaces of couture, as some historians have dubbed them (Myzelev, 2017; Palmer, 2001).

The luxury designer brands grew out of the entrepreneurship of few designers who left the fashion houses to start their own labels to pursue product differentiation. Most famous is perhaps Yves Saint Laurent who left the house of *Dior* to cofound his namesake label in 1961, exploring the options of merging haute couture with street sensibility (English, 2013). The high-end fashion houses did not see the need to do that because their customers – ladies of society – had little social need to be smartly dressed on city streets. In Saint Laurent footsteps followed Gorgio Armani, who freelanced for Italian fashion houses until he founded his own label in 1975, focused on high-end men's fashions – something the fashion houses deemed unimportant (Potvin, 2017). Men were supposed to be dressed properly, not fashionably. Similarly, Gianni Versace left *Genny* to open his own luxury boutique in 1978, to push the envelope with designs overtly celebrating sexual liberation, in contrast to the core definition of fashion style, which was to be subdued and lady-like. He is famously quoted by explaining his nonchalance when the Italian high fashion scene shunned him early on by dismissing the relevance of his work with the statement: "…Armani [Versace refers to him as the epitome of Italian elegance at the time] designs for the wife; Versace designs for the mistress… (Bilyaeu, 2018)."

The unique feature of these style innovators – the founders of the luxury designer brand – was an understanding of social change. This understanding provided the opportunities to explore product differentiation to serve fashion buyers neglected by the fashion houses. The fashion houses, especially the Parisian ones, did not bother with the needs of buyers who were not ultra-rich (Wenting & Frenken, 2011). As Crane (1999) explains, they set the trends to last only for a year, building their empires on selling a new outfit to each customer each fashion season. A high-end fashion buyer would purchase at least one new "ensemble" in the latest style each fall and each spring. By the time those styles diffused to the department store customers, as it took time to copy, manufacture, distribute, and promote them, the ultra-rich would be wearing something different. In social science, this dynamic was named the "trickle effect" (Fallers, 1954).

The fashion houses stayed on the forefront of trendsetting by establishing a system of design superiority for social elites that would trickle down to ordinary citizens. Their reputations were built and cemented by the fact that they only hired the best designers. They promoted that fact to their exclusive clientele, lauding the creative genius of their style gurus.

This culture of high fashion was centered in the European legacy of fashion's aristocratic birth. It was being challenged in Europe by design innovators following the social changes post World War II, but it had also become challenged by American design and retail innovations. It was in America that the department store business model emerged and as a function of it, the advent of what is known today as "activewear," "casual wear," "business wear," or, generally in fashion parlance, "*American* sportswear" (Martin, 1998).

American sportswear got its name in the beginning of the century as the type of elegant, but relaxed style of dressing, worn at leisure outings, centered around the watching of spectator sports (Goodrum, 2015). These were product categories for the American middle-class woman (Whitaker, 2006). American designers started to gain popularity as the American middle-class woman became the core customer of post-World War II fashion economics.

Fashion history posits that American design formally entered the world stage of fashion prestige with the Battle of Versailles Fashion Show on November 28, 1973 (Draper, 2015; Givhan, 2015). The chronicle of this event won Givhan (2015) a Pulitzer prize for detailed analysis of the cultural changes it represented. Spearheaded by Eleanor Lambert, legendary American stylist and founder of New York Fashion week, the show was a fashion competition between a group of the five most-celebrated French designers – Yves Stain Laurent, Pierre Cardin, Emanuel Ungaro, Christian Dior, and Huber de Givenchy, and the five most commercially successful designers from the United States, personally picked by Lambert – Oscar de la Renta, Steven Burrows, Halston, Bill Blass, and Anne Klein.

Anne Klein – the best-selling of all the American designers at the time – faced a strong objection from the French contingent because she made exclusively "sportswear" and not "fashion" items like evening and ball gowns. The other American designers had a strong focus on eveningwear and were known for their gowns and dresses. Anne Klein's empire was grown selling pants, pantsuits, swim and athletic wear. Eleanor Lambert was unwavering to the objection of the French contingency at Versailles, arguing that Klein should showcase why her label is redefining the concept of fashionable style for items other than evening gowns. Anne Klein opened the show with a collection of swimsuit type items with a tribal African motif, something that shocked the French. Assisting Klein that night was young Donna Karan, later to become known as the "Queen of 7th Avenue." Karan describes the battle of Versailles as the formative event in her professional development (DuVerney, 2016). It showed Karan the future of fashion design – a future not of exclusivity but of utility, functionality, and diversity.

In front of an audience of 700 guests, which included the most visible cultural and economic personages of the time – European industrial elites and royalty, American Hollywood stars, and art and popular culture icons – the American designers brought the guest to their feet in applauding a show featuring models of color, disco and funk music, strobe light visuals, and designs celebrating casual comfort. It was in stark contrast to the elaborately staged but cumbersome couture show from the French designers who did what they had always done – present high-end exclusivity for society's elites. The American designers presented their vision of how the modern everyday woman dressed and also redefined who that modern everyday woman was.

In cultural history the event is discussed for its impact on critical race theory because of the choice of models of color the American designers brought with them to Paris. But the choice was not pre-meditated (Draper, 2015). The girls were really favorite models of the designers. They were not chosen to go to Paris for the Battle of Versailles to make a racial diversity statement. They had established themselves as top models in America, reflecting a social change that was happening in the industry in the United States. The designers brought them to Paris because of their reputation as the best runway models in New York – the fashion capital of America at the time.

The American models had also created a unique runway style of walking from the legacy of being in shows set to popular music, favored in American fashion show production. It was a catwalk with elements of dance and exuberant movement, which designers encouraged in order to highlight the ease of movement of the clothes. The American style of design was first and foremost about comfort. Showing how easily the models moved in the clothes was essential. To the French designers and European audience during the Battle of Versailles show, that was a revolutionary concept. Seeing its power on stage was undeniable and in judging, the American designers were deemed the winning team.

The event was a turning point in fashion's hierarchical history because it was then that the European design royalty admitted defeat to the creativity of the American designers, allowing for the notion that fashion trends could be started not just outside of Paris, but outside of the influences of the social elites, as they were still understood at the time to be ultra-rich whites. The black models represented more than the racial history of American society. They represented a market. They represented the evolution of the American industry toward product differentiation for socio-economic strata with different levels of wealth, different cultural identities, different lifestyles, and therefore, different style needs. Those style needs were of the American working woman. *Vogue* fashion editor Anna Wintour explains that in the 1970s "the whole world started to look at the American working woman and what she was wearing...What she symbolized... and the role of women... All eyes switched to the United States..." Wintour concludes, meaning the figurative eyes of the fashion industry (Bailey & Barbato, 2012).

Fashion historians mark the Battle of Versailles as the turning point that allowed New York to become, what is known as, a "fashion capital." From this point on, the sales of American brands in Europe started a trajectory of such growth that during the 1980s would even lead to a cultural backlash in France. There, legislation was passed to ban the sales of blue jeans – the quintessential innovation of American fashion (Miller & Woodward, 2012). American brands grew in popularity on the European market precisely for the reason Anne Klein was almost excluded from the Battle of Versailles competition – their focus on sportswear.

Since then, New York fashion week has grown in importance to become one of the core fashion weeks in the follow-up stage of the industry's evolution – the creation of the global fashion brand. America was the place where the "designer as the brand" business model evolved in contrast to the "fashion house as the brand" European legacy.

The American designers at Versailles had started their namesake labels. In their footsteps, global mega brands of American activewear emerged from the 1980s on – *Calvin Klein, Ralph Lauren, Donna Karan* to name a few – that perfected the strategy of assigning luxury status to non-luxury product lines. The luxury status was the brand name itself. Building brand awareness became a cornerstone of marketing strategy not just in fashion and related goods, but across industrial sectors.

In fashion, consumers paid for the brand represented by a logo – the ultimate visual symbol of design. The iconography of branded commerce is subject to much fashion, cultural, and economic analysis, which examines it from a critical perspective of promoting social injustice (Klein, 1999) to a pedagogical tool perspective in the analysis of best practices of marketing strategy (Kim, 2005). Whole disciplines in business and service management are devoted to marketing strategies in branded commerce. They all focus on relationship of price and brand strength. The emblem of brand strength is the logo.

The visibility of a logo builds "brand awareness," as is the term, meaning socially popular based on a reputation for quality and excellence. The strong the brand awareness, the higher the prices of a fashion label's product lines. Therefore, this branded retail business model was in itself subject to a competition on price.

Fashion labels had to spend a lot of money directly (advertising) and indirectly (public relations) to build and protect their brand reputations in order to defend charging relatively high prices. Such a strategy works in national markets, but in the global market – comprised of different socio-demographic segments with different purchasing power spending abilities – success requires the aptitude to sell to multiple market segments that do not have luxury product spending habits or capability. This fact led to the proliferation of "brand extension" commerce (Choi et al., 2011; Colucci, Montaguti, & Lago, 2008; Dewsnap & Hart, 2004).

Once again, the main business model innovators were the American luxury brand designers. *Ralph Lauren, Donna Karan, Calvin Klein, Tommy Hilfiger*, and others grew to global mega brands status by differentiating product "down-stream," in terms of price levels. They retailed high-end luxury lines, and lower-priced "prestige" and "masstige" lines, actually giving birth to the concepts of "prestige" and "masstige" (Truong, McColl, & Kitchen, 2009), meaning product branded with a luxury logo, but sold at price points between luxury and mass-market levels.

Also called "category extension," this strategy of product innovation was quintessential to global brand proliferation (Choi et al., 2011). Specifically, Keller (2003) explains that in the 1990s, 80% of new fashion products were positioned via category extension. It is now an established business strategy in branded fashion retail to rely on the masstige product lines as most important for profits, as they tend to reach the buyers who strongly value the brand, but are also price-conscious (Paul, 2015).

It was this evolution of downward-spiraling price competition that merged the branded retail with the value retail business models. The value retailers – brands without a superstar designer's name as a logo that did not participate in fashion week – were learning from the designer label conglomerates that if brand awareness is the key factor for the successful retail of apparel, they could build their own brands with the promotion of apparel features, not designer personalities. Those that were successful in doing so – the *GAP, Abercrombie and Fitch, American Eagle* – appeared next to the designer labels and co-existed in the marketplace as trend followers.

This evolution of fashion economics, from the French-centered fashion house style dictation to the American-perfected strategy of global brand proliferation, happened while the system for the actual creation of style remained largely unchanged from the beginning of the century. Then and in a way to this day the world's leading designers show collections twice a year – in fall for fashions to be worn the following spring, and in spring for fashions to be worn starting the following fall. Two seasons, as is the term, where a handful of designers showed collections of options they posited would be popular six months into the future.

During those shows, the apparel producers evaluated the design statements for feasibility of what they promised – to be popular in the future – and also for their ability to be translated into clothing for the general consumer. High-end designers suggested trend ideas, but it was the apparel industry as a whole that proliferated the trends. With the onset of globalization, it was the apparel producers and retailers who diffused fashion, as Crane (1999) puts it, downward to the everyday buyer.

The fashion industry bifurcated into the commerce of fashion and the commerce of clothes. The commerce of fashion catered to its traditional market of exclusive luxury product buyers. With globalization that market became increasingly international, with new customers to serve outside of Western Europe and the United States. Fashion commerce expanded with the focus to increase sales internationally by differentiating product lines in non-apparel categories, selling the brand's image (Bridson & Evans, 2004). The apparel sector copied the trends the fashion sector dictated and retailed them with its own, independent stream of commercial innovations.

The main force in that stream was product differentiation toward more affordable lines, creating the very strong sub-sector of value retail, which gradually became, the "cash cow" of the industry, accounting for the majority of its profitability (Ross & Harradine, 2010). Its growth was based on a single logical economic truth – you can sell more product at low prices. It was the learning process of how to do so consistently from year to year, while competing to build brand awareness that moved apparel commerce into its own sphere of product positioning and marketing innovations. The marketing innovation trends evolved on two fronts. One was the advent and growth of global branding, giving rise to the term "global brand proliferation" (Hollis, 2008; Moore, Fernie, & Burt, 2000; Palumbo & Herbig, 2000). The other is the growing focus on the youth market (Azevedo & Farhangmehr, 2005).

Global branding research has shown that there is a cultural connection that people form with certain well-known and well-regarded brands that are external to their own nations (Alden, Steenkamp, & Batra, 1999; Batra et al., 2000; Steenkamp, Batra, & Alden, 2003). The acquisition, possession, and consumption of products from such brands build an imagined global identity. Therefore, global brand commerce became especially popular in the industrializing developing world with segments of the population who felt socially and economically excluded from the relatively higher standards of living in the West (Bartsch et al., 2016; Guo, 2013; Strizhakova, Coulter & Price, 2008).

Global branding has become the international cultural emblem signifying the relationship between self-identity and social-identity. Consumers use global brands as displays of social symbolism to build their self-image as a branded identity that links them to a "social identity" of a global citizen (Elliott & Wattanasuwan, 1998). When in different countries consumers use the same brands, convergence follows. It is described as "convergence of tastes and preferences," or generally as "cultural convergence" (Bikhchandani, Hirshleifer, & Welch, 1992; Lieberson, 1993).

In fashion, Godart and Mears (2009) describe the phenomenon as a convergence of "collective taste." Convergence is a process driven by integrated international trade that builds, as Douglas, Samuel, and Nijssen (2001) term it "international brand architecture." The building of an international brand architecture is the coordinated proliferation of multiple products under the same brand that are simultaneously positioned in multiple international markets. The goal is to build such a level of consumer confidence in these brands as to thin out local cultural characteristics and replace them with a unified, global convergence in tastes and preferences.

Cultural convergence has been so prominent under globalization and with the advent of new media that it is studied across the academy. It is the focus of research not only in fashion economics but also in popular culture diffusion trends, global governance, and soft power (Drezner, 2001; Gans, 2008; Sriramesh & Vercic, 2003). Cultural convergence is a direct outcome of the success of global branding.

As a marketing strategy for the fashion sector, global branding evolved from an economic need to actually minimize product differentiation of design features because implementing a differentiation strategy is costly. In the years after the Battle of Versailles, as American designers became popular in Europe, they established a strategy of differentiating product lines in each European market slightly to better respond to local tastes (Rantisi, 2004; Wigley, Moore, & Birtwistle, 2005). International product differentiation gained impetus in the 1980s and became the standard of, as is the term, "international product positioning" – the selling of different product lines at different price points in different countries – for clothes, as well as accessories and home goods (Markham & Cangelosi, 1999; McGoldrick, 1998).

The problem was that it was expensive to produce the high variety of different goods, to coordinate their manufacturing and distribution linkages, and to promote their differentiation features in traditional fashion advertising print and media campaigns. Therefore, as media became globalized in the 1990s, fashion advertising embraced the opportunity to dictate uniform looks of style and fashion in different countries, without having to offer local changes. Hence, started the period of "global branding" (Cayla, & Arnould, 2008; De Mooij & Hofstede, 2010; Roberts & Cayla, 2009)

During this time of media globalization and cultural convergence, fashion audiences in the three strategic main markets – North America, Europe, and Japan – were influenced by the new global entertainment culture, which since the 1980s had become increasingly dominated by American programming (Grainge, 2007). From this platform, American brands were able to set style internationally, and lower their global costs of style differentiation. At the same time, they pioneered and perfected the core retailing competitive strategy of the industry – price differentiation of congruent design but sold at different price points.

The reason why this tactic revolutionized retailing is the fact that engaging in competition to offer branded, but more affordable options, changed the main fashion market demographic. Those more affordable options were particularly popular with younger consumers. Competing for them eventually followed the same downward spiral as competing on price in the advent of prestige and masstige merchandising.

Competing on price means consistently looking for ways to offer lower and lower prices. Competing for younger customers means consistently managing to reach younger and younger buyers. The identification of this fact was the third factor that revolutionized the industry. The building of a production and marketing industry to cater to young buyers led to the birth of fast fashion.

Until fast fashion's rise – from the 1970s refocus of the fashion designers to serve the working woman, to the 1990s redirection of keeping her stylish by offering branded but more affordable options that she can change more often – the core fashion customer had always been assumed to be a female, aged 18–32 (Blaszczyk, 2011). Some of the early fashion marketing research roughly defines that age demographic as "29 and younger" or "college-age women" (Schrank & Lois Gilmore, 1973; Summers 1970).

This acceptance that "college-age women" were the main customer demographic defined the legacy of the field. That bracket was studied for all its consumption attributes – from disposable income to the ability to act as the fashion "opinion leaders" for all apparel-buying segments (Goldsmith, Heitmeyer, & Freiden, 1991; Polegato & Wall, 1980). The age bracket below it – 14–17 – was not included in major marketing research until the proliferation of value retail in the 1990s (Auty & Elliott, 1998). Today, it is this fashion segment that drives fast fashion's meteoric profits. Long discounted by fashion merchandising, it is the teen market that is the core of the industry's economy.

Value retail evolved around branded retail to address the teen demographic, gradually becoming more important with the impetus of global youth culture. Value retailers were the precursors to the fast fashion retailers. It was their successful operational strategies in catering to the youth market that the fast fashion retailers perfected. Understanding why the refocus occurred is essential because, just like the racial diversity statement the American designers made at the Battle of Versailles, it was a product of cultural change – the elevation of the teenager to the status of a cultural icon.

Again, America was the innovator of the creation of the teenager as a cultural archetype in terms of fashion and style. With Hollywood showcasing the lifestyle of an American teenager with the films of John Hughes in the 1980s for example, including such iconographic titles as *Pretty in Pink* and *Sixteen Candles* and teenage pop-stars such as Tiffany, Debbie Gibson, and New Kids on the Block, conquering the pop music world, the dressing of teenagers and showcasing their obsession of how they look, birthed a retailing refocus toward the youth market.

That focus stayed with the industry, which learned that youth-focused value retail was less prone to economic uncertainty. Even the theory of a "lipstick effect" emerged analyzing why during periods of economic downturns, certain fashion-related item sales – cheaper merchandise and cosmetics, hence the name "lipstick effect" – actually increase (Hill et al., 2012; Netchaeva & Rees, 2016). This is particularly true in emerging markets where the Western value brands were and still are expanding rapidly (Dickson et al., 2004; O'Cass & Siahtiri, 2013). Hence, youth brands become more focused on style in building competitive advantage (Azevedo & Farhangmehr, 2005).

It became clear that sales are less elastic at price points that young customers can afford. In order to compete with other retailers for those sales, as the competition could not be based on lowering prices further, youth retailers refocused on promoting style. The core of these three differences between the adult and teen (and younger) markets is in, as is the term, "consumption function" of the youth market. Consumption function is the economic concept of customer-buying behavior as a function of spending ability, in the context of choice for substitute goods (Friedman, 2018).

In fashion economics, understanding the consumption functions of different market segments is essential for effective advertising. The whole point of fashion, and all other advertising, is to change the "elasticity" of the consumption functions of buyers. In economic terms, elasticity is the willingness to switch among product options and/or not make a purchase at all, when one is evaluating the price and attributes of a product (Hartmann, 2006). In the context of fashion, the race to build brand awareness, create brand allegiance, and promote a brand's core attributes is done with the goal to lower the propensity of its customers to buy different brands and/or curtail their purchasing habits when faced with economic hardship.

In traditional fashion promotion, branding had been successful as the leading strategy to lower the elasticity of the consumption function of fashion-conscious buyers. Yet branding can only go so far when the economic conditions change during downturns and recessions. Therefore, it is well established that luxury and higher-priced fashion product lines are notoriously susceptible to economic recessions (Allenby, Jen, & Leone, 1996; Browning & Crossley, 2000; Reyneke, Sorokáčová, & Pitt, 2012).

The advent of value retail showed market analysts that it was significantly less prone to economic contraction during recessions – a key discovery that would redefine the basic concepts of fashion's *mass market* and *core consumer*. In the mature fashion markets of Western Europe, Japan, and the United States, trend analyses of the proliferation of value retail consistently indicated that it was economically much more resilient in terms of not only sales but also brand expansion (Hampson & McGoldrick, 2013; Kumar, 2007). The popularity of value labels such as the *GAP* and *Old Navy* was growing not just among youth markets but also among older buyers.

Value retail is less prone to economic contraction than the retail of high-end-priced merchandise because during recessions, customers make sacrifices on big-ticket items, not low-priced goods (Shipchandler, 1982). Fashion consumers in particular become more value conscious and search for more affordable options. Value retail was particularly effective in the rapidly industrializing Asian nations, known as the Asian Tigers. It offered local buyers American and Western brands and the cultural prestige of their ownership, without the luxury good price tags (Arvidsson, 2006; O'Cass & Siahtiri, 2013; Stephen Parker, Hermans, & Schaefer, 2004). However, value retail faced one major challenge for further expansion – it did not set trends. The problem was that value retailers were not style creators. They followed the style dictates of the fashion labels. The innovation that changed this dynamic and catapulted fast fashion brands into the forefront of style trendsetting was, one more time, a cultural as well as economic change – the advent of cyber society.

The Advent of Fast Fashion

In political economy it is said that a market change is caused by a cultural change (Mantzavinos, 2004). The cultural change that dethroned the European fashion houses as the world's style dictators was the growth of America's cultural power. American culture embraced diversity and that fact is reflected throughout the history of the evolution of American fashion and its retailing.

American apparel retail rests on a legacy of promoting images of social change. From the entry of American fashion on the world stage to the evolution of the way it was sold and to whom, the American retail experience set the standard for global operations. It was the American labels that promoted apparel through treating it as an emblem that was reflective of cultural change. The legacy is an understanding that the promotion of the celebration of cultural change is best received by the young.

Fashion became refocused from the lady of society toward youth culture. The attention and promotion to younger consumers evolved with the proliferation of "mall culture" and the technological innovations of media that increased the ways through which producers could engage young consumers. This increase in producer–consumer exchange led also to opportunities for youth culture to influence and keep redefining fashion style.

Fast fashion allowed youth culture to have a voice in the creation of style. New media provided the channels for that voice to travel through the veins of the industry in ways different both in kind and in type. The dissemination of style changed from the legacy of designer-created and fashion-press dictated, to consumer-defined.

Two factors fuel this change to today. One is visual media, and specifically "user-generated content" (Daugherty, Eastin, & Bright, 2008; Fischer, 2011; Van Dijck, 2009; Wyroll, 2014). The other is "integrated marketing" (Duncan & Everett, 1993; Schultz, 1992). In their simplest definitions, or rather examples of the concepts, user-generated content is the use of blogs, online communities, and specifically in fashion, *Instagram* posts. Integrated marketing, having been subject to some discourse in the 1980s and 1990s in terms of definition, today is best summarized as the combination of all traditional promotional tactics – advertising, sales promotion, public relations, and direct response to consumer input, into an integrated "message" (Broderick & Pickton, 2005). Therefore, it is often referred to as integrated marketing communication (IMC) (Holm, 2006; Lane Keller, 2001).

Kliatchko (2005) tracks the evolution of IMC definitions and explains that in the digital world, the early definitions of "direct response" – from the late 1980s through the 1990s – fail to capture the nature of today's reality of cyber information participation from byers. In modern commerce, when each producer has an online retail option, there is also an online comments and customer input channel. Information from the consumers on these platforms has become essential in product development and promotion. Therefore, it is the advent and use of both user-generated content and IMC in tandem that has created a platform of disruption in style trendsetting. Understanding the impact of their interplay is important because it defines the modern evolution of fashion retail.

Codifying and defining the different forms of user-generated content, Wyroll (2014) makes the critical observation that its various forms change the commercial model of information transmission. The author explains that society has moved from a consumer-oriented communication culture to a communication culture of participation. This observation is based on the analysis of social media commercialization of Fischer (2011), who puts the change from unidirectional to user-generated content in the context of professional specialization.

Unidirectional content is created by professionals. In fashion, those are the fashion editors, critics, and fashion journalists who covered the collections, provided analysis of the trends on display, and choose options. They used to dictate style. The designers proposed style; the professional fashion press accepted or rejected it. User-generated content on the other hand is customer driven. In blogs, cyber community platforms, "haul" videos, and social media forums shoppers offer their personal voices and points of view on products and brands.

Commercially, this egalitarian cyber participation brought about first the "fashion blogger" and now the "fashion influencer" (Sudha & Sheena, 2017; Wigley, Moore, & Birtwistle, 2005). Influencer content defines fashion analysis today on par, or often despite and contrary, to the professional fashion press (Bendoni, 2017).

With the growth of digital user-generated content, research on integrated marketing abounds, analyzing its transformative impact on traditional advertising. For example, Schivinski and Dabrowski (2016) explain that integrated marketing allows consumers "direct voice" into the product development process, which is a positive factor for both consumers and producers. The research on "brand transference," meaning consumer awareness of brand features that builds brand loyalty, notes that today this "direct voice" increasingly promotes messages of social justice and equity (Acharya & Rahman, 2016; Lock & Harris, 1996). Consumers prefer brands that back social causes such as environmental stewardship, human and animal rights, as well as show evidence of corporate engagement in general social justice activism. The messaging is overwhelmingly transmitted via new media on smart devices.

The advent of smart devices changed fashion branding through the online communities first created by fashion bloggers and now defined by fashion influencers. These trendsetters and style leaders promote looks rather than labels. They mix and match items from different labels and price points, promoting value based on novelty – the ability to quickly change looks. They also promote options to make purchases fast, on which online commerce relies. The simultaneous growth of cyber culture and cyber commerce link in the fashion industry in their reliance on the core factor that is behind the transition to consumer-defined style. That core factor is low prices.

Frequent purchases are the fuel behind a business model of promoting novelty. Novelty relies on fast change. Fast change can best be achieved when buyers do not worry about affordability. Most importantly, this business model can be profitable only if customers also do not worry about the brand power of products.

From innovation in materials and design to changing trends, looks, and behavior, success in mainstream fashion commerce is dependent on the ability of producers to put "product on shelves," as is still the commercial term, at mass-market prices. Although today much of apparel retail happens online and, in terms of percent-change growth, that trend is expected to continue around the globe (Nguyen, de Leeuw, & Dullaert, 2018; Srinivasan, 2015), the retail culture of the industry remains vested in the traditional model of retailing in a physical space. Fashion sales are still defined by what's on shelves, racks, and on mannequins in stores. Store spaces are filled with options.

To offer such options, as already explained, brands manufacture multiple product lines to sell simultaneously at budget, mass, masstige, and prestige price points – the core fact behind the concept of "brand extension" (Liu & Choi, 2009; Forney, Joo Park, & Brandon, 2005). In modern day retail, which is a hybrid of physical store and online commerce portals, customers can shop for clothes at each of those price point levels from the same brand. Luxury retail remains vested in luxury pricing, yet even there, some brands have begun offering lower-priced merchandise. For example, *Versace* has *Versus, Armani* has *Armani Exchange*, *Emporio*, and *Touché*, *Prada* has *Miu Miu, Gucci* has *Aspiration*.

When faced with many options, the modern fashion buyer values low prices above all else (McColl & Moore, 2011). Therefore, the industry's overall profitability relies on product lines sold at the lowest price points. This access to affordability and choice is the change agent of fashion retail. Through the evolution of business innovations in value retail from the 1970s to the onset of fast fashion in the early 2000s, the fashion industry changed its retail model from one of top-down style diffusion, as Crane (1999) tracks it, to one of bottom-up trendsetting.

This bottom-up trendsetting culture lies on the brand power of the fast fashion labels. What is unique about the success of the power of their brand strength is the fact that it is devoid of a personal face. There are no big-name designers associated with *H&M* or *Zara*'s product line launches. They use celebrity models in promotion, but do not showcase specific designers as a manifestation of style creativity. Fast fashion has established that style is secondary to price. Furthermore, it established that branding can also be built on price, or rather, on the promotion of low price as a key branding attribute. When consumer today sees the logos of *Zara* and *H&M* – the two largest retailers in the world – they see brands with prestige and style power without question. Two decades of promoting the low prices of the brand, rather than its designs have changed the way fashion retailers build brand awareness.

Chapter 2

The Fast Fashion Paradox

The Culture of Immediacy

Fashion is an active part of culture (Crewe, 2017). In today's culture of visual immediacy, meaning fashion customers are themselves visual brands existing on social media platforms, it is a cultural concept of promoting a wishful state of being. Cyber activities such as *Instagram* posts, *Facebook* and *Twitter* updates, or *TikTok* videos allow each fashion enthusiast a unique platform of creative self-expression. What is being created, however, is an exaggeration of personal aspirations, mixed with fantasy. This user-generated way of marketing creates far-fetched fantasies around the meaning of what we buy (Crewe, 2017; Vehmas et al. 2018). Such fantasies – from the beginning of haute couture to today's fast fashion reality – are the core of what successful fashion commerce is. It is the sale of illusions.

While in the past, our illusions of who we wish to be were carefully dictated and sold to us, today we are the creators of our own illusions. We craft the image of who we wish to be through our clothes. The more mystery, mystique, and intrigue, the better. Mystery is even part of brand promotion. The 2016 documentary *Zara: The Story of the World's Richest Man* by the Prime Entertainment Group describes how the founder of the famed brand, Amancio Ortega, carefully crafted the furtive fairytale story of his success by remaining mysterious. He would not do what is expected of brand owners and managers – engage in personal promotion. As a matter of fact, the film explains, Ortega had not made any public appearances or official interviews from 1975 until the ending credits of the film show him emerging from an elevator in 2015 at his 80s birthday party.

The culture of mystery has been promulgated through the evolution of *Inditex* to build a fashion empire devoid of the key components of traditional fashion success – designers and brand owner showmanship. The documentary tracks hordes of devoted customers, all musing that they have no idea how *Zara* does what it does, but how much they love the brand. They love it precisely because of this supernatural, almost magically unbelievable ability to provide whatever it is we were looking for, or had no idea we were looking for, when walking into its stores.

Or don't we have an idea? Of course we do. The above-mentioned documentary and plenty of other media stories abound with open accusations of plagiarism, copycatting, stealing designs, then mass-producing fast, and at scale, entire product lines based on stolen designs. *Zara* deploys hordes of trend watchers and analysts to scout popular designs, then skillfully appropriates their main features and offers clothes cheaply and forcefully. The firm never publicly addressed all formal accusations of design plagiarism – which means, it has also never officially denied it.

Without their own designer power, fast fashion conglomerates openly steal the designs of others. All have been accused; most shrug off the accusations (Cohen, 2012). They call it "democratization of fashion" and "bringing fashion to the people," but by open theft of intellectual property? Frankly, yes. Because once a design is out there, and it is being promoted on its own merit, it can be copied without much legal repercussion (Felice, 2011).

The combination of different designs is a similar matter because there is a difference between repeating and imitating. In order to stop a designer, or a producer such as *Zara*, from repeating a particular design, creators often copyright specific design features (Martin, 2019). Few visible examples are the red sole shoes of *Christian Louboutin* or the pineapple prints of *Stella McCartney*. Yet, imitating or creating very similar looks is the norm in fashion.

Fashion keeps repeating itself. That is the history of fashion. For example, when asked by iconic fashion journalist Jeanne Beker about his ability to dictate fashion trends, Karl Lagerfeld famously said back in the early 1990s that he does not dictate. He "proposes" a trend and then waits to see if others will adopt it and make it successful. What is different with the fast fashion model is the speed of production of a look from "proposed" to "rack," as is the industrial expression.

In effect, the economic history of fashion commerce is based on imitation. However, historically it took much longer for a trend to both be commercialized and proliferated. It took up to a year to be able to sell garments with certain design features in the global market in a context similar to today's Zara-style of retail. From being shown in a collection, to having the collection pieces readied for mass-production, often meaning even changing the designs a bit to make them more commercial, to placing the "spec" orders with sub-contracting factories, to those factories being able to source the right materials and actually manufacture and arrange for the shipping of garments to retailers, it took an average of 18 months (Arnarson & Hardarson, 2014). During that time, fashion magazines, television and visual media campaigns, and other traditional marketing platforms, actively promoted "the trend" of the moment. All this promotion was done before the clothes were on the shelves of stores, much less having had the opportunity to be worn and evaluated by consumers. Then the feedback from customers, mainly measured through the volume of sales, would actually show if a particular design or a trend really reached popularity.

Modern technology allows for items to be mass-produced and positioned in weeks, with no need of all these steps. It is based on, as is the new term actually coined by *Zara's* public relations promotion, "sourcing" of ideas. In traditional fashion parlance, the term "sourcing" had been used to refer to selecting producers, as in "global sourcing." Now it describes the sourcing of trends. The competition is to offer them with lightning speed. Trendsetting now is discussed in the context of "micro trends" – lasting a month to three months (Arnarson & Hardarson, 2014).

Even at this rapidly changing trendsetting pace, the core aspects of traditional fashion trend-setting features remain. Retailers such as *Zara* and *H&M* compete on selling "trendy" clothing. Yet, their business model is based on a competitive strategy not to set the trends, but to follow the trends. It is now taught as part of corporate strategy in business schools, with *Zara* being credited as the creator of the "trend-following" retail model (Ghemawat, Nueno, & Dailey, 2003; Caro & Gallien, 2012). However, as explained in Chapter 1, that is not entirely the case. Value retail was based on this very natural commercial method of mass-producing items that followed a trend inspired by the creations of legitimate, professional designers. Fast fashion merchandising is just following the basics of value retail but at a much larger scale and much faster turnover. Now, super-fast fashion does it with a much wider breadth of trends.

Super-fast fashion is the newest reincarnation of value retail. Super-fast fashion allows for the creation of thousands of styles, aligned with the latest trends (Cachon & Swinney, 2011; Fletcher 2010). It is the new way that leads emerging fashion innovators into favorable competitive positions. For example, fairly newcomer to the fashion seen, and fast growing in popularity with celebrities and customers, *Misguided*, describes its business model as "rapid fashion," introducing around 1000 new styles a week. It is this tactic of aligning multiple styles with a one or a few core trends that is driving the success of the new crop of emerging apparel retailers. They have learned from the fast fashion leaders – *Zara, H&M, Primark, Top Shop, Forever 21*. Some fast fashion powerhouses are already struggling to compete with the new innovators. *Forever 21* filed for bankruptcy in 2019.

The Super-Fast Fashion Innovators

The new "kids on the block" are thriving due to one main retail competitive advantage – no physical stores. *Forever 21* relied on its stores in shopping malls as the main selling platform. The new game changers are *online mainly*, or increasingly (and only to become more significant in the aftermath of COVID-19) *online only* retailers. They are also emerging globally, many of them focused on serving the vast and economically growing markets of the Global South.

The fast fashion conglomerates, although relying on online commerce, still pursue a much targeted retail strategy to position flag-ship stores in exclusive retail hot spots. It is estimated that *Zara* paid $300,000 million for its New York City location on Manhattan's 5th Avenue (FilmsMedia Group, 2016). The super-fast fashion retailers have no such expenses. As they have no or very little retail overhead cost, they can achieve two business goals that help with building competitive advantage. The first goal is the ability to retail at low prices, and in the case of super-fast fashion, at even lower prices than the *Zaras* and *H&Ms*. The second goal is the ability to shift much of the final manufacturing functions locally to Western nations, saving time on shipping expenditures from low-cost production locales.

Those innovators come mainly from the United Kingdom – a nation with a strong fast fashion culture and much studied in apparel consumer behavior. The UK is both the epitome of fast fashion commerce as well as sustainable fashion activism. The paradox is that such activism and the growth of super-fast fashion seem to exist separately in the British social consciousness. Both are gaining impetus at the same time, having polar opposite messages. The reasons for this paradox are cultural.

Arguably the most successful British super-fast fashion brands *Missguided, Boohoo*, and *PrettyLittleThing* are headquartered in Manchester – a historical textile and clothing production city that suffered economic decline with the advent of global value chains (GVCs, a concept to be discussed in Chapter 5), when most manufacturing relocated to Southeast Asia. The growth of these new conglomerates should be carefully tracked as they are pioneering a fairly new way of localizing production, also referred to as "reshoring" (Gray et al., 2013; Rashid & Barnes, 2017). Localizing production is a goal in sustainability because it has implications for "decarbonization," as is the emerging term (Bini & Bellucci, 2020), of apparel supply chains. They are among the longest, and most carbon-intensive, as cargo is shipped repeatedly back and forth between the developed and developing world, as explained in detail in Chapter 5 of Anguelov (2015).

The new super-fast fashion brands source fabrics and trim from close-by (in geographic terms), yet still fairly low-cost (in terms of integrated pricing), nations in Eastern Europe. The process and its history will be explained in some detail in Chapter 4. The fact that the super-fast fashion brands also manufacture locally, and in the case of *Missguided, Boohoo*, and *PrettyLittleThing*, in Manchester, UK, allow the producers to benefit from a political platform of showcasing economic patriotism. Perhaps for this reason, beneficially branding the fact of "doing the right thing" economically for their local communities, super-fast fashion brands are not admonished for mass-retailing at super-low prices.

The success of these brands creates a venue for social and political support of their business model. Embracing it can lead to expansion in local employment, addressing the needs of many historic industrial hubs in Western nations that have lost apparel manufacturing jobs as a function of global value chain economics. Localizing or "reshoring" manufacturing is going to be a major political goal. Political rhetoric of economic patriotism is growing in the developed world, and firms that can capitalize on its popularity can gain both economic and political clout.

The promotional tactics of the super-fast fashion brands are to rely on influencers and *Instagram* celebrities, to earn them that coveted social status of "cool" labels. This promotion is based on showcasing constant change in designs and looks. Their teenage customers adopt a "look" as "hot" for a minute. But for their next *Instagram* photo, a new outfit is the ultimate goal.

The super-fast fashion labels emerged and flourish online. Online commerce allows them considerable savings on up-front costs and physical retail space expenditures, as well as high degree of agility in distribution, as saving can be redeployed into an expansion of distribution infrastructure. That tactic allows for customers to receive items quickly, increasing their propensity to make more purchases online. Young buyers cannot resist these facts, fueling super-fast fashion commerce and culture. It is a segment that is "constantly shopping" through multiple devices, clicking "add to cart" with ease, impulsively, and regularly.

Enter the Influencers

Shaping young people's personal style needs is the influencer culture. Fashion influencers are the promotional platform of today. They are the new trend setters, style creators, and in general, fashion innovators because they mix and match clothing items, looks, accessories, and most importantly, brands. Additionally, they are a unique economic engine because their "job," to put it directly, is to generate Internet traffic.

The emerging economic term is "click-bait" commerce, and unfortunately much academic attention is devoted to its deployment in political misinformation and cybercrime (Kirwan, Fullwood, & Rooney, 2018; Vultee et al., 2020). Generating click-baits relies on, as Pengnate (2019) puts it, "emotional arousal." It is largely based on images, with trigger words as "reinforces." In terms of the way fashion brands use click-bait platforms, McKelvey (2019) offers the example of British fast fashion retailer *All Saints*. *All Saints* uses the services of Los Angeles-based "digital organizing software provider" *NationBuilder* (precise self-description from *NationBuilder's* website) to connect with its online customers via enticing, politically charged messaging, especially when introducing new product lines.

In terms of generating huge volumes of exposure, click-baiting is so successfully used by fashion influencers that a new term has emerged to describe the entrepreneurial success of those who make it – "Insta-famous" (Boerman, 2020). There is fierce competition for that status and once achieved, an Insta-famous influencer is courted by firms that want to benefit from his/her/their promotions. This fact generates even more power, as the influencer followers attribute expertise and style knowledge to popular influencers, elevating their fashion expert status.

As number of followers and amount of brand recognition increases, so does the power of an influencer. McFarlane and Samsioe (2020), using a netnography approach to examine the features of hundreds of popular fashion *Instagram* posts, find that the majority come from around 50 influencers. This is a concentration of style decision-making power akin to the traditional fashion-week-based model, but removed from it. The authors find that influencers refrain from directly posting endorsements of specific fashion brands, but rely on emojis and hashtags, in a type of "post-construction mix of messaging." The findings also show evidence of attracting followers with politically charged posts. Yet, there is no mention in this study, or others that analyze the political messaging in click-baits, of sustainability and/or environmentalism.

Influencer research is linked to the works that analyze the importance of individuality. The modern fashion buyer values uniqueness above all. A culture of fashion choice has formed based on creating a unique look, generally using a mix-and-match approach, combining multiple brands, vintage items, and even consignment pieces with, unfortunately, fast fashion items (Cervellon, Carey, & Harms, 2012; Giovannini, Xu, & Thomas, 2015; Halvorsen, 2019; Phau & Lo, 2004). The influencer click-bait platform drives the speed with which this unique-look culture changes "looks." It is daily. Therefore, this culture conditions consumers to feel a need for a constant stream of new items. Unless one is very wealthy, this constant stream of new items would need to be fairly cheap.

In this current culture of cyber promotion, based on user-generated content, fashion brand awareness and brand prestige become secondary to stylization influence. Style is not dictated by professionally trained stylists, critics, and trend-setters, but by consumers with active social media presence. It is a reality of immediate consumer response, as is the professional term. It allows producers to see how their main customer base responds to a new product via social media. There, in public forums that are global, customers and their influencer leaders, engage in product evaluation and critique. Sales figures reflect the outcome also immediately, allowing retailers unprecedented ability for response. User-generated content, as opposed to traditional advertising, now defines not only fashion, but all on-online commerce (Müller & Christandl, 2019; Smith, Fischer, & Yongjian, 2012; Tirunillai & Tellis, 2012).

The effectiveness of user-generated content is immediate and volatile. In the fluid cyber platforms of blogs, *YouTube* channels, *Instagram* posts, and *TikTok* videos, non-professionals share input, and in that way, it is the public that influences fashion, rather than the other way around. Yet, even with this high level of consumer participation, trend-setting has a hierarchy. Although all of us, as consumers, can be part of user-generated communication, we do not participate equally. Our personal cyber presence varies and this variety allows for influencer power to grow.

Influencers follow the familiar pattern of creating looks of fashion stylists, which links brands in what is known in traditional fashion parlance as "brand adjacency" (Reddy et al., 2009). Both today's influencers and traditional fashion stylists pick and choose items from different brands and showcase the ensembles as new looks. What is different today is the fact that up until the influencer era, brand adjacency was created and dictated by the brands themselves. The fashion stylists mainly took cues from the brands. What is also very different is the cultural and national identity of not just the influencers but the new conglomerates, which increasingly come from non-Western nations and represent a non-Western cultural aesthetic.

Stars from the East

As Chapter 5 is to explain, it is the consumption in such nations, spearheaded most notably by China, South Korea, and India, which drives the global fast fashion market growth. Where in the beginning of the fast fashion era, the expansion was driven by Western brands, today firms from the East are aggressively competing. They include brands that start local, but expand in the global market, such as *Uniqlo, Zaful, BAPE, Charles & Keith, Pomelo Fashion*, and *Lady M.* These are Asian brands with Asian cultural identities, proudly not reliant on Western fashion legacies or culture.

The power of the global expansion of Eastern cultural exports, which include those fashion brands, can be exemplified by the success of Hallyu – the Korean Wave (Kim, 2019b). Park (2011) states that it is simply the popularity of Korean entertainment in other countries. However, it is not that simple to define the phenomenon of Hallyu because it would be reductive to equate its scope to entertainment value. It is a socio-cultural impetus that is behind the proliferation of Korean entertainment because it shows the evolution of Korean culture into a modernity of post-border, digital internationalization. This internationalization is carefully nurtured by the Korean government, which has an official industrial policy for promoting Korean *style* abroad (Kim, 2017; Park, 2011).

According to an assessment from leading international management consulting giant Globeone[1], Hallyu commerce is built on (a) integrated marketing, as most of the promotion relies on user-generated content of virtual social networks, blogs, and platforms devoted to entertainers; (b) promotion by "celebrity models" – for example, K-pop stars and actors are their own issue-brands of social justice causes, becoming "cultural ambassadors," as *Globeone* puts it; and (c) building of a strong international *brand architecture* – products are carefully positioned in complementarity to each other, offering product lifestyle lines based on South Korean tech exports. Factors (a), (b), and (c) save for the South Korean tech export specification, are perfectly describing the tenants of the fast fashion and super-fast fashion business models.

The international expansion of Hallyu started regionally in South East Asia as early as the 1970s (Oh, 2016). Once it proliferated strongly in Japan in the early 2000s, Hallyu became a popularly used term internationally. Japan is the main Asian cultural power and its own cultural exports, such as anime for example, have been growing in popularity in the global entertainment market since the 1980s. Still, until the mid-2000s, the Japanese, and the general Asian entertainment market, continued to be dominated by Hollywood and Western European fashion, popular music, and programming. In the 2000s, increasing amounts of South East Asian entertainment products appeared.

The largest Asian nations started to export their national entertainment to global audiences. From Bollywood musicals to Chinese martial arts blockbuster movies to entire TV channels being dedicated to Japanese animation appearing the world over, culturally based entertainment products became important industrial export sectors for emerging market nations. Among them, Korean K-pop music and TV soap operas, in particular, reigned supreme in popularity all over Asia, with their presence in Western markets increasing at a rapid rate. Eventually, even Hollywood paid homage to the popularity of South Korean entertainment by awarding the 2020 Oscar for best picture to the South Korean black comedy thriller "Parasite."

Park (2011) explains that some of the main components of this success lie in fashion, in particular, challenging regional cultural stereotypes of gender relations via powerful fashion statements of culturally transformative imaging. The female characters in TV, film, and online programs celebrate high emancipation and independence, in relation to general Asian cultural norms, dressing provocatively, if not ostentatiously. The female K-pop groups celebrate unabashed sexuality in their fashion choices, challenging the Asian cultural extolling of modesty. The male characters of TV and film are romantic, genteel, unapologetically handsome, tailored in dress, and carefully groomed. Although not many of the analyses equate this image to the Western phenomenon of metrosexuality, there are strong similarities. They challenge the strong authoritarian archetypes of cultural masculinity of Korea, Japan, and China. Hwang (2009) gives a detailed account of the reception of Korean actors in the region, stating that in Japan they are more popular than Japanese actors because of their metrosexual image.

In terms of fashion trend-setting and influence, Shin and Koh (2020) explain that the Korean wave style is, at large, "genderless." The K-pop boy groups are endogenous and increasingly non-regional in image and style, exhibiting high influence of African American hip hop, Caribbean reggae, and American "emo" sounds and looks. The impact is visually mesmerizing. This optical connection is explored in a series of articles for CNBC in 2012 where non-Korean speaking, mainly Asian fans of K-pop were interviewed.[2] Most accounts explain that it was the visual component of the music that first caught their attention, usually through YouTube channels. It was the videos that make it essential to understand the concepts of the songs, since the fans do not understand Korean.

In all such accounts, academic works, journalistic exploits, and business analyses one recurrent theme emerges as an explanatory factor of Hallyu's popularity – visual impact. One would expect the appeal to be connected to the technological prowess of Korean industrial innovation such as highly sophisticated special effects, transformative design, surprising color schemes, or the promotion of new materials, textures, and products. To an extent that is the case with all successful merchandising of visual product. However, cross-cultural communication experts who closely study the cultural content of Hallyu strategies explain that the visual appeal is mainly based on transformative cultural messages (Hwang, 2009; Jin & Yoon, 2016; Park, 2011).

Hallyu is the definition of a modern cultural export because it captures the dynamics of the social mediascape of Korean culture (Jin & Yoon, 2016). This fact is exemplified in the promotional tactics of Hallyu stars. Although they are unmistakably and proudly Korean, actors, singers, and fashion celebrities proudly exhibit their knowledge of Japanese and Chinese language and culture. What Korean Hallyu stars also do well is the "new normal" in celebrity self-promotion, which is to expand their marketability across entertainment sectors. Singers act, actors sing, and many launch fashion, fragrance, and home décor lines (Kim et al., 2013). This entertainment sector fluency generates celebrity-defined commercial activity where the celebrity is the brand, building his or her own brand equity.

Celebrity as *promotion* is much studied in the brand proliferation and management fields because of emotional connections. The incarnation of this fact in fashion advertising lies in the simple example of who is on the covers of *Vogue* magazine. Since the late 1990s, this most-coveted and iconic fashion image has been of a celebrity, not a professional fashion model (Hall, 2000).

Researchers study consumer brand allegiance to celebrities because of the social justice advocacy of "celebrity spokes models," as is now the phrase (Kelting & Rice, 2013; Okonkwo, 2007; Partzsch, 2015). Celebrities pick certain issues, building their own "goodness" image. Celebrity political activism, broadcasted through the advertising campaigns of the brands they promote, is a major focus of digital commerce research, referred to as "issue branding" (Brockington 2014; Daley 2013; Ponte & Richey, 2014).

Issue branding – the promotion of a social justice issue a firm supports, rather than simply the quality of its products – has become "the tactic" to build a positive brand image in the eye of the consumer. Kelting and Rice (2013) explain that consumers remember the celebrity rather than the advertised products and transfer their feelings about the "goodness" of the celebrity to the brands, mainly associating them with the image of the celebrity. This dynamic is referred to as "integrated marketing" – a two-way promotional partnership. Okonkwo (2007) explains that in successful integrated marketing, the celebrity model or spokesperson must have global appeal. The end result is global branding not only of products but also of public figures, and the social justice causes they support. For example, Kim (2019a) notes the emergence of sustainability messaging from certain K-pop stars on fashion waste, which the author describes as "eco-criticism" of Hallyu entertainers in general. Hallyu stars aggressively promote their self-based brands via "self-fashioning," and in that way act as style influencers. Few of their own colleagues have become critical of this fact. In their art they include social messaging against eco-centrism, promoting recycling. In a related study of fast fashion avoidance in Korea, Yoon, Lee, and Choo (2020) find evidence of anti-consumption beliefs in their sample of female consumers, aged 20 to 39, as a function of personal environmental impact concerns.

Such research is emerging, yet is not as prolific as the works on fashion sustainability initiatives in the West, which are to be tracked in detail in Chapter 3. How the East and Global South markets will react to the industry's call for the transformation of business practices remains to be seen. As of yet, most research on Asian fashion economics is focused on popularity, style differences, and international market growth (Cheang & Kramer, 2017; Jin & Cedrola, 2016).

With respect to Korean fashion, Park (2011) analyzes the fashion statements of Hallyu stars and their international success. The author explains that the success of Korean style is in no small part due to active support by the Korean government and strong collaboration between the government and Korean companies. In relation, Kim (2017) analyzes these government–fashion industry partnerships and expounds that government agencies, fashion associations, which are also government-supported and funded, as well as fashion companies and their related-industry collaborators, all play an active role in creating and promoting Korean style. In other words, in South Korea, the exporting of fashion style is an industrial policy. The government provides policy support for sectors with products that can best be promoted via celebrity. Park (2011) explains that the goal of this collaboration is to develop an interest in Korean culture, with the goal to sell more Korean products overseas.

It is working well. People who become fans of Korean celebrities pay attention to the products those celebrities favor and become fans of Korean products, food, culture, and fashion. Hwang (2009) finds that international consumers had formed strong attachment to Hallyu as a brand because of a perception of high quality. In that study "brand" is defined as a celebrity, TV show, or a film. The author states that consumers "actually purchase more Korean products, especially if they learn that the actor, actress or singer they love uses the product, and even visit Korean more often (Hwang, 2009: 206)."

This fortuitous outcome of the promotion of cultural exports strengthens the national brand of South Korea. The popularity of South Korean products increases the economic power of the nation, and it defines its diplomatic clout. The influence is, no doubt, strongest in the region and as the geopolitical power of South East Asia increases, so would South Korea's importance as a global power. Such an outcome is the ultimate goal of successfully nation branding.

Nation branding is a concept defined by British marketing executive Simon Anholt and developed theoretically with the proliferation of globalization (Anholt, 2006, 2007). Anholt tracks the development of the concept from inception to achieving international recognition in 2005 when the New York Times included its annual list of new ideas. What is unique about the idea is that it acknowledges the importance of the public sector.

Nation branding makes the distinct connection between diplomacy and the promotion of multinational brands. Nation branding is more than international awareness of Italian food, or Indian yoga, or Belgian chocolate. These concepts are termed "cultural exports." Nation branding takes the impact of the international success of cultural exports and levies it on the image and reputation of a nation's government. Kaneva (2011) offers a thorough chronology of nation branding research, examples, and specifics, making the clarifications that unlike diplomacy, and even most forms of soft power, nation branding is built on the opinions of consumers. It is, as Kaneva (2011: 118) defines it: "reconstituting nationhood through marketing and branding paradigms."

Linking this vein of research that tracks the interplay between culture, branding of cultural and national governance, cultural convergence, and its impact on consumption and global marketing, a common factor is evident. It is its reliance on user-generated content. As Ash (2016: 7) puts it:

> Never in human history was there such a chance for freedom of expression. If we have Internet access, any one of us can publish almost anything we like and potentially reach an audience of millions.

This is the nature of user-generated content. All of us can publish anything, respond to anything, start communities, and influence communication development. In fashion, this fact is the bases of commerce and business innovation. Professional content is indistinguishable from user-generated content, and even professional brand advertising is reliant on click-bait promotion.

In this user-generated, multi-cultural, cyber-fashion democratic forum, a paradox has emerged. As the volume of fashion sales escalates, there is growing advocacy to reuse and recycle clothes (Ekström & Salomonson, 2014). There is major problem with such advocacy, to be explained in detail in the following chapters, which is the fact that recycling clothes is still not commercially viable, or ecologically safe. Yet, the concept is there and innovations are underway. Still, as such "reduce-reuse-recycle" initiatives and movements, including the "slow fashion" movement, emerge (Ozdamar Ertekin & Atik, 2015; Pookulangara & Shephard, 2013); disposable clothing sales are growing at a steady rate (Remy et al., 2016).

A comprehensive report released by the Ellen MacCarther Foundation in 2017 shows that apparel sale volumes globally continue to rise, while the number of times a clothing item is worn decreases significantly (Ellen MacCarther Foundation, 2017). With over 600 sources, the report provides concrete data on the amount of clothes purchased globally, with alarming estimates of the subsequent environmental outcomes. It contains contributions from some of the biggest global brands, including *H&M*, *Adidas, Patagonia*, and *Tommy Hilfiger*.

The report urges readers to change their ways of obsessively buying cheap clothes and to make better choices, of course highlighting the technologies (many of them to be outlined in this book) underway, which would offer less-toxic options. Yet, without an honest discussion of the higher prices such innovation would require, the reader is left with just a political polemic of how things should be changed, but little guidance on how to change his/her/their own behavior. Except, of course, the familiar mantra – buy fewer and more expensive items.

This message is vacuous. It has been shown in over a decade of research that consumers will not opt for more expensive and "better" choices when cheaper substitution options abound. Bick, Halsey, and Ekenga (2018) explain that despite all the environmental stewardship social messaging, the retail culture of the industry encourages consumers to devalue their clothing.

Since, as already explained, the core consumers are teenagers, researchers have started to analyze the buying habits of the youth demographic with a focus on sustainability. They are finding that even in cultural locales where climate change and sustainability concerns have become social norms – mainly Western Europe, the United States, and Japan – young peoples' fashion consumption does not reflect such concerns (Barnes, Lea-Greenwood, & Joergens, 2006). Specifically, Chi (2015) explains that consumers confess to care about social and environmental factors, yet they become secondary considerations, as the primary influence of buying behavior is the price of a garment. Vehmas et al.'s (2018) results also indicate the leading factors in apparel buying to still be the price and quality of an item. Sustainable inputs or information about ecological stewardship is just a welcome reinforcer, if present. In relation, Joy et al. (2012) find that for young consumers, sustainability or supply chain social justice problems are not priorities when making shopping decisions. What is important is meeting their own personal style needs. The authors also find that even if some consumers practice sustainable consumer behavior with certain goods, they do not include apparel items. In other words, it is very common today for young (and old) to recycle, buy organic produce, use public transportation, bicycle, or walk more to decrease their carbon footprint, yet they continue to buy fast fashion.

The steady and continuous rise in aggregate fashion sales is a testament that consumers are ignoring, dismissing, or simply not acting via their buying behavior on environmental stewardship messaging (Remy et al., 2016). Cyber-fashion still promotes rapid-fashion products, even in light of ecological and sustainability advocacy. The reasons are linked to the propensity of click-bait communication to spread misinformation. In fashion communication, this propensity has been embraced by the entire fashion culture to create another fairly new terms with roots in fashion that now have been applied to other industrial products – "greenwashing."

Notes

1 See http://globe-one.com/power-of-culture-hallyu-the-korean-wave-4636/
2 For example, see Naidu-Ghelani, Rajeshni. (July 16, 2012). "Move Over Beiber - Korean Pop Music Goes Global." cnbc.com http://www.cnbc.com/id/48157880

Chapter 3

The Changing Face of Fast Fashion

Challenging Fast Fashion

It has been investigative journalism that has raised the visibility of fashion's ecological crisis. Among the most important reports are Rosenthal (2007), Cline (2012, 2018), Zarroli (2013), and more recently Chen (2018), Danigelis (2018), and Shanghani (2018). Offering a thorough account of the ecological consequences of the consumption of cheap clothes, Cline's (2012) *Overdressed: The Shockingly High Costs of Cheap Clothes* has become both an international best seller and has been academically cited hundreds of times – a testament to the scholarly interest on the issue. Among the most noted early works on the environmental damage of the industry, Luz Claudio's analysis from 2007 stands out. It is among the most-cited and discussed articles because it links the pollution of the industry's growth to issues of public health. Titled "*Waste couture: Environmental impact of the clothing industry*," the article explains that the main problem is the promotion of a wasteful consumer culture (Claudio, 2007).

From an activist perspective, the Greenpeace "Detox" campaign, which was launched in 2011, played a powerful role in bringing the ecological impact of fashion manufacturing into the business consciousness of brands. The early days of the campaign pitted *Nike* and *Adidas* against one another to see which would first accept Greenpeace's challenge to "detox" their manufacturing processes. *Puma*, owned by Rudolph Dassler, the estranged brother of *Adidas* owner Adolf Dassler (hence the name and European pronunciation

Adi-Das) took the opportunity to leap out in front and pick up the "Detox" torch. Dassler, determined to not be outdone by his estranged brother, took "the bait" and the three largest global activewear brands joined the sustainability quest. Today, *Adidas* is leading in product innovation, announcing the development of "The Loop" – a fully recycled shoe made from reclaimed ocean plastic, employing a fusing technology to replace glue. James Carnes, vice president of Brand Strategy for *Adidas* states on the episode "Killer Kicks" of the BBC's documentary series *Newsbeat*, "… almost 100% of all shoes in the sports industry are glued… and glue is basically poison for recycling…" (BBC, 2020). The Loop is expected to come on the market in 2021.

Adolf Dassler's action back in 2011 provided the Greenpeace's "Detox" campaign with legitimacy. Its goal of eliminating the discharge of hazardous chemicals in fashion production entered a real phase of commercial activity. Big-name brands joined the campaign, often after intense, aggressive, consumer-mobilizing actions in or in front of their shops and on their social media sites (Greenpeace International, 2012).

The momentum of social media-led activism grew and brands were forced to address the impact of chemical use from a completely different perspective than the previous focus on consumer safety. The earliest consumer safety policies had to do with limiting or banning chemical components that could leach toxic elements on the body of consumers. For example, one of the best-known consumer-safety certification bodies, OEKO-TEX, founded in 1992, has developed several certification standards, for harmfulness when worn. The main three standards are Oeko-Tex 100 certification, the Oeko-Tex 1000, and the Oeko-Tex STEP.

After the Greenpeace "Detox" campaign, rather than simply thinking about consumer safety when wearing or caring for garments, fashion conglomerates had to address workers' safety while manufacturing their products, and the exposure of manufacturing communities to hazardous chemicals through the discharge of contaminated water. This was new territory for most brands, and certainly not something buyers were prepared to think about.

Then, and in a way too today, fashion-conscious buyers were faced with ever-increasing variety of incentives to keep buying relatively cheap clothes. This rate of increased purchasing is the economic growth engine of the entire industry. Therefore, its promotion is the lifeline of the sector. Only very recently has a criticism of the ethics of the promotion of overconsumption in fashion started to increase. Academically, research lagged behind, while investigative journalism took an urgency stance with articles, editorials, and in 2015, the first full-length feature documentary films.

In 2015, the documentary *The True Cost* generated high visibility of the environmental and labor exploitation problems in apparel commerce as a consequence of the proliferation of fast fashion (Ross & Morgan, 2015). The media picked up on the popularity of the film with contributions of celebrities such as Livia Firth, wife of actor Collin Firth. Congruently, talk show host John Oliver dedicated an episode on his HBO show *Last Week Tonight with John Oliver*, titled "Fashion" (HBO, 2015). That year, in 2015, the advocacy spotlight was on fashion sustainability, in part spurred by the growing volume of criticism by social justice scholars and activists who posit that the fast fashion business model rests on cheap labor (Bhaduri & Ha-Brookshire, 2011) and it is this reliance on labor and environmental exploitation that defines its financial success, while fueling global poverty, inequality, and pollution (Plank, Rossi, & Staritz, 2012). It was at the end of 2015 that the prequel to this book was published (Anguelov, 2015).

Although there was research and academic work on the ecological issues of fashion economics (Chen & Burns, 2006; Claudio, 2007; Morgan & Birtwistle, 2009), it was not until the tragic factory floor collapse in 2013 at the Rana Plaza complex in Bangladesh that the media started to examine the problems in an integrated, systemic way. The documentary film *The World According to H&M* is a prime example of such systemic investigation (Maurice & Hermann, 2017). It links evidence of labor and environmental exploitation by tracking the profits at each link of the international production network of *H&M*. Examining the corporate response strategy of *H&M*, the investigative team visits the production facilities of the recently launched Conscious Collection to show that it is no more environmentally or labor friendly than any other fast fashion product line.

With a more directed focus on environmental justice, the documentary film *River Blue* looks at water pollution from finishing and dyeing of fabric (Williams & McIlvride, 2017). Its conclusion can best be summarized by the words of Sunita Narain, activist and director of the India-based Society for Environmental Communications, quoted in the film: "We are committing 'hydrocide,'" meaning killing waterways by polluting them to hopeless levels. Two other documentary films build on the environmental damage from water pollution in fiber production: *Fashion's Dirty Secrets* (Langstaff & Onono, 2018) and *The Price of Fast Fashion* (BBC, 2018).

The topics of these productions vary from exposing labor exploitation, to unethical marketing, to environmental damage. Impact on the environment is the core problem, yet aspects are stressed differently. Also, the conclusions of the works offer different solutions or suggest divergent paths toward solutions of the main issues. In *The World According to H&M* the focus is on labor and financial exploitation. The solution implications are to increase transparency in governance and oversight of labor protection standards and to implement changes in compensation and taxation structures. The goal of the film is to show the disproportionate financial profits accruing to *H&M* from not paying a fair share in wages or taxes.

In *The True Cost*, labor issues are again the main theme, with pollution coming in second. However, water pollution is barely mentioned while the focus is on pesticide pollution from fertilizing cotton. With that focus, unfortunately, the sustainability solution is not really a solution at all – it is to favor, support, and switch all cotton production to growing organic cotton. Defaulting to organic cotton as the recommendation, however, is a major oversimplification of a complex scenario. The problem is that organic cotton is much more water intensive to grow and in nations racing to capture global market shares as organic cotton exporters, ecological issues arise from soil quality degradation and watershed depletion (Dowd, 2008; Uygur, 2017). For these reasons, a new call has been initiated in cotton production – to support "better cotton" – grown with organic inputs, but also incorporating traditional agricultural practices (Zulfiqar & Thapa, 2016).

The Better Cotton Initiative (BCI) was born to address the challenge of organic cotton production's inability to scale up. For years, the percentage of organic cotton grown globally has hovered around the 1% mark. According to the Organic Trade Association, in 2019 a mere.07% of all cotton produced was certified organic (OTA, 2019). The costs associated with certification are a barrier, as well as the 2–3 year transition period (depending on the selected certification system), where farmers must meet the organic standards but cannot yet market their cotton as "organic."

The risk of such transitioning is substantial. If a pest or disease or weather issue poses an existential threat to the crop, BCI mandates permit the limited use of certain pesticides, insecticides or fertilizers. Organic mandates do not. BCI aims to educate farmers on the best techniques, using minimal inorganic inputs, and helping them to minimize costs, maximize yields and profits, and help build the economic power of their local communities. It is also currently at about 30% market share, significantly surpassing the organic market.

One of the main issue initiatives like BCI are trying to mitigate is water used during crop irrigation. This problem is well covered in In *Fashion's Dirty Secrets*, where Stacey Dooley – a well-known British TV personality – travels the world to uncover the hidden costs of "the addiction to fast fashion," as the promo copy of the film puts it (Langstaff & Onono, 2018). The film astutely analyzes the overall ecological impact of the production of clothes and identifies the biggest issues to be water use and pollution.

In the other documentaries, the main foci have mirrored previously well-established problems in the industry – labor exploitation, capital flight, pesticide pollution. Even *River Blue* does not offer the nuanced analysis of the industry's reliance on the use of large quantities of fresh water with such detail as in the data presented by Stacy Dooley. *River Blue* discusses overall water pollution mainly from denim and leather processing – the two largest "finishing," as is the term, textile manufacturing operations – mainly in China and for good reason. It is estimated that over 70% of all fresh watersheds in China are dangerously polluted by textile manufacturing (Pan et al., 2008; Webber, 2017). *Fashion's Dirty Secret* covers not only water pollution from the finishing treatments of yarns and fabric, but also from integrated water use in all stages of production.

Through *Fashion's Dirty Secret*, Miss Dooley accosts shoppers outside fast fashion stores, asks them to show their purchases, and then tells them how many liters of water were polluted in the production of the purchased items. Then the feed shows graphics of rows of 1-liter water bottles. The journalist would also add an approximation of the equivalency of years-worth of drinking water needed to sustain one person. The reaction of the shoppers is journalistic gold – shock, disbelief, remorse – all providing for an emotional viewer experience.

The other films, as well as the volumes of academic journal articles, popular press editorials, and books on fashion pollution, pale in comparison to the clarity of the numbers given by Stacy Dooley. In the film, she also tracks the human impact when the pollution of communal waterways is put in the context of the local populations that rely on them for daily hydration and irrigation. The damage of the large-scale irrigation for growing cotton is also powerfully illustrated by showing how the Aral Sea, which used to be one of the largest bodies of fresh water in the world, is now reduced to a dust plane, camels wandering alongside the skeletons of rusted out fishing boats. And this heavy dependency on water does not end once the cotton is picked.

Defining Fashion Pollution

Depending on what the cotton is destined to become, it could go through a variety of different processes, many of which rely on water. First, cotton must be scoured clean and bleached before being spun into yarn. Then, whether dyed as a fiber bundle, in the yarn stage, in the fabric stage, or even once it is made into a complete garment, the dyeing process depends on large quantities of heated water. After dyeing, the "finishing" stages can potentially be the most water-intensive, particularly for a product like denim jeans. Many processes used to soften, age, fade, add coatings, top colors, or prints, will require water in addition to significant amounts of energy and chemicals – including potentially hazardous ones.

Water use is a concern for other materials as well, beyond cotton. It is essential in irrigation at the agrarian stage in the growing of cotton, but also for flax for linen, or trees for cellulosics, and it is of course necessary for the raising of animals whose skin or fur are destined for the fashion industry. Afterwards, water becomes the life blood of textile production. Traveling from the Aral Sea in Kazakhstan to the Citarum River in Indonesia – a regional textile manufacturing cluster with over 400 factories – the investigative team of *Fashion's Dirty Secret* shows that around the world, but mainly in investment-starved developing nations, water pollution from fashion production is escalating at alarming rates. In the first stage of water use in the production of clothes, referred to in industrial jargon as the "agrarian" stage, millions of gallons of clean water are being siphoned and diverted from vital rivers and lakes for the growing of cotton. The film gives the following statistic: it takes 3,400 gallons in irrigation to grow enough cotton to make a single pair of jeans.

There is a reason why clusters of textile factories, such as the ones shown by the afore-mentioned films in Indonesia and China, are built on the banks of rivers. The rivers' water flow is used in two ways. One is as power – water is diverted through canals into the facilities to power the looms – and the other is in the treatment of yarns, fabrics, and garments in the employed "wet processes."

The main wet processes are scouring and bleaching, desizing, mercerizing, dyeing, garment finishing, laundry, printing, and coating. Each is reliant on large quantities of water, which results in the production of liquid effluents with varying waste composition of alkali, starch, acid, base, and bleach (Chen & Burns, 2006; Gordon & Hsieh, 2006). Dyeing is particularly problematic because colorfast dyes – meaning resistant to fading in repeated washing and caring for garments – may contain high concentrations of the carcinogens Mercury, Lead, and Cadmium and the heavy metals Zinc, Copper, Iron and Manganese (Dey & Islam, 2015; Kant, 2012; Molla & Khan, 2018).

Of all the wet processes, dyeing is the most toxic. Analyzing pollution from dyeing over a decade ago, Ibrahim (2008) argues that, by that time, dyeing technology was so inefficient that over 15% of the world's *total* production of *all* dyes was lost during dyeing of fabrics. It is a result of the necessary repeated dyeing and rinsing needed to achieve proper colorfastness and color precision, as well as remove unfixed dyestuff prior to consumer laundering. It is still true today that conventional reactive dye – the most commonly used dyestuff for cotton and cellulosics, which is used on about 70% of the world's cotton products – achieves fixation rates of only about 85%, leaving 15% of the dyestuff to be rinsed away. This is a result of the fact that cotton simply is not good at absorbing dye, and while there are now processes available to enhance cotton's affinity to dyestuff and improve dye uptake (cationic pre-treatments, for example), these products have not penetrated the market to the degree many had hoped.

The good news is that innovations in dyeing technologies, whether in terms of chemistry, machinery, or new materials, are developing rapidly. High fixation or polyfunctional reactive dyes for cotton, for example, have been developed in the last two decades. They achieve 90% or more fixation rates compared to the typical 75%, but they are more expensive (even though they save time and reduce auxiliary chemicals) and they take more skill to use (Khatri et al., 2015).

The combination of all the wet processes happens at large volumes using river streams, when strong and plentiful. When water scarcity becomes an issue, factories drill into groundwater wells, piping fresh water out (Angelis-Dimakis, Alexandratou & Balzarini, 2016). In Bangladesh specifically, groundwater extraction has become a real problem. As textile production is the main industry of Bangladesh, which Hossain, Sarker, and Khan (2018) estimate to continue to grow significantly, every year factories have to drill deeper to pull out the water they need, which has a devastating impact on communities. A local market has developed where the water itself has become the currency because factories pay for its extraction (Bhattacharjee et al., 2019).

As groundwater reservoirs shrink, their chemical concentrations change and there is a literature examining the high levels of arsenic found in village wells in Bangladesh (Alam et al., 2002; Paul & De, 2000). The arsenic spike is linked to water shortages caused by over-drilling groundwater wells, as well as by direct pollution from effluent runoffs. Such issues result from the high amounts of water that is needed to manufacture garments. On average 200 tons of water is used for the production of one ton of textiles (Greer, Keane, & Lin, 2010). The used, chemically laden effluents travel from the aquafers around textile plants into the ground water systems of large regions, affecting the toxicity of entire ecosystems (Ibrahim et al., 2008).

These facts all stem from one simple, inconvenient truth: textile pollution cannot be removed with existing filtration methods and clean-up technologies. Kant (2012) explains that out of the 72 toxic elements emitted in modern day textile production, 34 cannot be treated at all with available purification technologies. The other 38 can be partially treated, mostly through zoning rather than removal, meaning allowing toxic effluents to be expelled into waters classified as "low risk" for human contact (Cohen, 2014; Narayanaswamy & Scott, 2001). Innovations to address this problem are underway, such as Zero-Liquid Discharge (ZLD) platforms, where biological treatment – use of bacteria, and chemical treatment – use of crystallization and reverse osmosis, are employed to solidify liquefied pollutants into inorganic waste and salts (Ali et al., 2016; Gronwall & Jonsson, 2017). Such technologies are emerging, but are not widely used.

Research has shown that the majority of the textile mills around the world do not have adequate arrangements to treat effluents before discharging them into an external drain (Khan et al., 2009; Pan et al., 2008; Rosenthal, 2007; Tüfekci, Sivri, & Toroz 2007). Many external drains expel into waterways that also serve as communal irrigation canals. These studies explain that purification technology is extremely limiting in terms of being capable of removing harmful substances from the discharged runoffs. What little filtration mitigation can be deployed is expensive and therefore, factory managers generally do not invest in filtration systems. Even where facilities appear to have a modern wastewater treatment plant, referred to as "effluent treatment plants" or ETPs, comparisons of their production volumes versus the capacity of their ETPs often show that it is impossible for them to be treating all their wastewater (Sakamoto et al., 2019). Additionally, facilities facing cash shortages or late payments from their clients will sometimes shut off the ETP to save energy and money. Furthermore, alarming evidence is emerging that under pressures to deal with the pollution, local governments are devising plans to divert effluents further away from the factories. Investigating for National Geographic, photographer Lu Guang shot images of industrial piping, pumping polluted effluents into open seas in China. For his work and activism, Guang has won international praise and acclaim and the scorn of the Chinese government. The New York Times reported him as officially missing December 2018, allegedly by his wife, having been apprehended by Chinese authorities (Pledge, 2018).

A similar dynamic is explained by Mohan et al. (2017) in the city of Tiruppur, India, which has approximately 729 dyeing factories. Although the local pollution control boards have implemented a ZLD policy and claim all factories are equipped with ETPs, in actuality they are coordinating with state boards on a benefit–cost analysis for construction piping that would dump all effluents into deep sea. Until then, factories discharge runoffs directly into community waterways.

Images of this fact are the first thing that pops up in "fashion sustainability" internet searches. The opening scene of *River Blue* shows magenta colored effluents flowing into a river in China, as fashion designer and activist Orsola de Castro states: "…there is a joke in China that you can tell the 'it' color of the season by looking at the color of the rivers…" (Webber, 2017). Later in the film, de Castro explains that, as activism around pollution monitoring has grown, factory managers have learned to hide their pollution. They lay longer pipes underground to divert effluent flows further away from facilities. In cluster areas where there are multiple factories, it is impossible to identify the source of pollution (Williams & McIlvride, 2017).

After being expelled in the runoff process, the untreated chemicals affect the quality of drinking and irrigation water for hundreds of miles, as they travel in groundwaters far beyond factory zones. Khan et al. (2009) provide evidence that water pollution from clusters of textile factories in Bangladesh has led to the displacement of whole traditional communities and the destruction of entire ecosystems. Similarly, water discharge figures in China for 2003 show that pollutants in that industry were forth among the worst in content and volume of all industrial effluents (Pan et al., 2008). It must be stressed at this point that since these works raised visibility of the problem over a decade ago, research and data reporting on textile water pollution at large stopped. It wasn't until Greenpeace launched the *Toxic Threads* series in 2012 that the issue again gained visibility.

The most current investigative reporting from Greenpeace stresses that the problem is growing at alarming rates. Data from tests conducted throughout the world show effluent toxicity continues to be high above legal or recommended levels (Greenpeace International, 2012, 2017, 2018). Those "recommended levels" are the guidelines provided by the World Trade Organization (WTO), to be explained in detail in Chapter 4 of this book.

The problem is that those standards, well established in effluent regulation and research across industries (Bhutiani, Varun, & Khushi, 2018; Hessel et al., 2007; Molla & Khan, 2018; Prasad et al., 2015), are mere guidelines. Each country is free to determine tolerable discharge levels, establish its own regulatory structure and implement oversight and enforcement measures (Chaturvedi & Nagpal, 2003). Evaluation of the outcomes in some of the nations most dependent on textile production – Turkey, Pakistan, India, and China – indicates that the guidelines are irrelevant as factories generally do not comply with them (Tüfekci, Sivri, & Toroz, 2007; Chaturvedi & Nagpal, 2003; Khan et al., 2009; Pan et al., 2008).

Factory managers consider investing in treatment technology to be a non-productive use of funds in an industry that struggles against strong cost pressures. According to Khan et al. (2009), in Bangladesh treatment is regularly below standards and is rarely checked either by the factory, environmental department, or buyer. Pan et al. (2008) report that in China, in particular, standards vary across regions because of centrally planned development policies and many local governments allow companies to emit waste beyond legal limits.

These findings are decades old and come from the most cited academic works that raised the visibility of the ecological damage of fashion production. Despite their impact, and the growing social awareness of the issue, more recent toxicity evaluations show an escalation of pollution emissions (Ghaly et al., 2014; Greenpeace International, 2017, 2018; Siuli, & Mondal, 2017; Watharkar et al., 2015). That is the case because economic incentives are missing for industry owners and managers to comply with what little regulation there is. As a result, the industry operates at very high pollution volumes associated with the manufacturing of a simple garment. For example, accosting one shopper outside an *H&M* store in the United Kingdom, Stacey Dooley explains that the production of just one button-down shirt, which the man had just purchased, had polluted the equivalency of three-years-worth of drinking water that a human consumes.

Yet, at large, such information is absent from the narrative the general public reads or hears on fashion sustainability. It is missing from most books on sustainable fashion activism and economics. Admittedly, the technicality of the textile production process is an issue, as is the variability of the statistics on the required inputs. Yarn, fabric, and garment manufacturing are industrial operations with specific production steps dependent upon specific chemistry and biochemistry knowledge. The assembly processes pose varying risks and hazards based on an almost infinite number of variables, ranging from region of production to product type and performance requirements.

All innovations in the sector have been driven by a quest to improve the quality of the final product – fabric. To that effect, the knowledge system that has been created is one of agrarian and industrial steps to create versatile, durable, fine, pliable, blend-able, and most importantly, affordable textiles. Knowing how to do that is a process of learning for industry professionals.

When non-industry professionals investigate and interview production and operations mangers about their craft, it is understood that textile production experts would have knowledge outside the level of expertise of others. It is perhaps this fact that is behind the evidence of lack of a knowledge base in textile production, also referred to scientifically as "textile mechanics" (Schwartz, 2008; Zhou, Sun, & Wang, 2004) by the journalists covering "sustainable" fashion. In very few of the articles, reports, and documentary films, is there viable discussion of the most toxic of production stages in fashion – wet processing, meaning bleaching, dyeing, and the industrial laundering steps of fabric. When mentioned, it is glossed over rather quickly and the focus, as is often in journalistic writing, turns to promoting solutions. It is this natural incentive in investigative reporting that allows the fashion conglomerate public relations representatives to capture the information flow.

Greenwashing with the "Sustainability" Slogan

When fashion journalists ask industry spokespersons about fashion sustainability, it is the industry's representatives' voices that are heard. They have their unwavering incentive to present the producers in the most favorable way, to promote their own firms, and to only provide information that is positive. And journalists need a positive message.

Doom and gloom reporting only goes so far. Readers are also consumers and they need feel-good endings in articles and clear solutions to the presented problems. Therefore, the majority of fashion pollution reporting touts recycling, buying organic product, and as of late, even limiting the laundering of clothes.

All such messaging is immaterial in the context of the most hazardous aspect of apparel production – wet processing. However, the soundbites are familiar to consumers and recognizable as something that consumers can immediately do. They can adjust their buying habits, unwaveringly looking for that organic logo; they can put their old clothes in the recycling bins; they can get a pair of *Levi's* "no wash" jeans (Levi, 2018). In terms of fashion consumer behavior, those are major purchasing habit changes. They feel like immensely important actions to consumers, and therefore, there is a tendency to equate making such actions with doing enough to offset fashion pollution.

Consumers cannot be expected to know details of the production process in any manufacturing, especially for goods that are made in complex petrochemical processes. Yet, it is this simple fact that is not conveyed to the fashion buyer – that clothes are petrochemical products, manufactured with acids, heavy metals, and dangerous carcinogens. Consumers also cannot be expected to know which products are recyclable and to what degree. For such knowledge, they trust the information presented to them from producers and those analyzing production.

The dissonance behind the misinformation in fashion sustainability that has even led to the call to limit the use of the term sustainability lies in the fact that fashion sustainability analysis that most directly reaches consumers comes from fashion journalists. The problem is that there is little evidence in the majority of works on fashion sustainability that the journalists know enough about fiber production and features, chemical processes of fabric treatment, or market incentives for recycling clothes, to ask probing questions of feasibility and veracity.

What exacerbates the problem is that fashion journalists ask industry representatives who have a strong incentive to promote their own eco-branding agenda, portraying their firms in the best possible light. For example, writing for the *Saturday Evening Post*, Nicholas Gilmore extensively quotes Jackie King, the executive director of Secondary Materials and Recycled Textiles (*SMRT*) – a company whose job it is to collect discarded fabric (Gilmore, 2018). Miss King states that 95% of used textiles can be recycled. That is an outright lie the journalist does not challenge. Miss King never explains how clothes can be recycled but does lament that in the United States only 10% in fact are, quoting estimates by the US Environmental Protection Agency (EPA) report (EPA, 2016), blaming it on the unwillingness of the consumer to properly dispose of unwanted clothes. Then the article actually gives an example of what *SMRT* can do, which is turn fabric into insulation and stuffing materials. That process is *not* recycling. Recycling is turning waste of a product back into product, or turning clothes waste into clothes. Using clothes waste as input in the manufacturing of other products is "downcycling," which is quite different from recycling (Cao et al., 2006). The output is of lower quality and lower commercial value than the input.

A related example comes from another documentary program by the BBC, which has launched a series of sustainable fashion themed news segments. The text on the BBC portal for the special hour-long report titled *The Price of Fast Fashion* clearly states that the investigative reporter – Assefeh Barrat – is a fashion lover (BBC, 2018). Miss Barrat is shown consistently in fashionable outfits as she travels from cotton farms in America to textile factories in Turkey to eco-fashion boutiques in New York City to "see who is leading the way in reducing fashion's environmental impact."

As journalists do, Miss Barrat goes in search of a narrative. Fashion PR reps are eagerly awaiting opportunities to offer a myriad of narratives. Journalists report those narratives, allowing for misinformation to spread. In the film Miss Barrat visits a factory complex in Konak, Turkey that claims to have implemented modern textile dyeing innovations toward sustainability by deploying computer technology to increase precision in color matching. While this can genuinely save significant water and energy otherwise expended during repeated dyeing cycles, it is simply good practice – not even best practice – and certainly not a worthwhile credential that denotes sustainable production. Yet to Miss Barrat, that is enough evidence, as she is shown nodding and smiling. Then the camera moves to Emirah Bilgin, Head of Manufacturing, Coats, who explains that in the dyeing process the factory "tries to use recycled water; not fresh water, to reduce waste." As Mister Bilgin speaks, the camera shows the effluent discharge shafts from where dirty effluents flow into a grassland. Abruptly the conversation ends with that image and moves to the next steps in the production chain – the weaving of fabric.

The problem is that this seems like the perfect solution and the piece of good news the journalist is seeking: "we are trying to use recycled water" but it is so vague that it could be completely false, yet it would remain unchallenged. A simple follow-up question: "Would that be possible?" would beg for an explanation, which could be obtained by asking a logical series of questions, as in: "What do you mean "recycled water?" From where? Wouldn't it be chemically laden in a way to change the color precision?" And even more technical questions should be asked in relation to particular trace elements left in waters from the pre-color treatment of yarns, such as alkali and starch, and how they impact color precision and subsequent quality measures such as colorfastness. However, this is never the case. We are yet to see a journalist who has asked any specific questions about coloring technology. Instead, when the journalist receives (in most cases) a well-rehearsed response, peppered with eco-friendly terms, the conversation moves on.

Next, the documentary film shows the marketing director of *Soktas*, a cotton fabric manufacturing firm headquartered in Konak, Turkey, explain how this global supplier works with hundreds of customers at a time to fill particular fabric specifications for each. For every order, the company has to source particular yarns. Often, under significant price pressures, the fabric supplier has to choose the cheapest option, which is not the best sustainability outcome. The factors that fabric weavers, such as *Soktas*, can control to help sustainability transparency are upgrade technologies to save on energy and water expenditures and use sophisticated computer technology in the finishing and quality control stages in order to reduce waste.

After this part, Miss Barrat launches the narrative we hear repeatedly – that there is a small revolution happening in the industry toward sustainability. A story we love to hear, but a story that does not represent the full truth, which is that the industry needs to be transformed, or it faces an existential crisis. The examples show only very small-scale, very expensively produced, boutique-type, high-end-designer-type goods, targeting eco-fashion-conscious consumers. While such businesses are emerging, they are not changing the industry. Fast fashion commerce surges on, greenwashed with the "recycling" slogan.

Taking up the charge of greenwashing promotion, in yet another BBC documentary *The Sustainability Challenge*, part of the series *Protecting Our Planet*, the investigative team is granted an interview of only one fast fashion retailer – *H&M* (BBC World News, 2019). Giorgina Waltier is introduced as the "head of sustainability." However, her official credentials on the company website indicate her role to be "sustainability manager for the UK and EU," for the past year, and before that a Corporate Press Officer. Before joining *H&M* Miss Waltier was an editor at *GQ* magazine.

There is a big difference in a "head of sustainability," "sustainability manager" for a region, or an official public relation's spokesperson, or "corporate press officers," as is the *H&M* definition under which Miss Waltier officially started her career with the company. The main difference is that "heads of sustainability" oversee the manufacturing of product lines. Corporate press officers do not. They are public relations' directors, heading the in-house advertising functions of a company. In the corporate social responsibility culture of activism promotion, these corporate spokes hubs are publicity entities. Their employees' and supervisors' skills are in communications, not in operations management. Their job is to promote an image of sustainability, not implement, oversee, or evaluate sustainability practices.

Again, it is the knowledge of the journalist that is possibly behind the mistake of identifying Miss Waltier with a title that has misleading implications. Yet, it is understandable if the corporation creates such official titles precisely because of their ambiguous meaning. When facing inquiries, such "heads of sustainability" can use the title's meaning to justify statements about operations they do not supervise.

Understanding production operations in global conglomerates such as *H&M* is not easy. One of the main reasons is the fact that they do not own their own factories. Tokatli (2008) lists the fast fashion retailers that do and those that do not. *H&M* does not. *Zara*'s corporate parent *Inditex* does, but not all.

As explained in Chapter 1, many apparel producers today, are "born global" meaning, they do not own any production facilities but rely on out-contractors. *H&M* is one such conglomerate that relies on independent producers through a series of "tiers" (Wada, 1992). Tier 1 suppliers often refer to the sewing facilities; this may be the extent of a brand's knowledge of who is in its supply chain. Tier 2 is likely to be wet processing, and Tier 3 may be fabric production or raw material suppliers.

Although brands have different ways of defining tiers, and everything shifts when there are agents involved that help broker deals between brands and manufacturers, none of the tier suppliers' clients have supervisory functions or powers. There are initiatives underway, and some research is examining their effectiveness. For example, M. Tachizawa and Yew Wong (2014) explain the evolution of sustainability mandates through the tier system and note that often first-tier suppliers train managers in lower-tiers to use environmental databases. In fashion, that process is more complicated than in many industries because manufacturing occurs in several stages, which are to be explained in detail in Chapter 4.

In the context of sustainability mandates, the process is evolving to include increasing number and variety of participants. A multitude of organizations, such as consultancies and non-governmental organizations (NGOs) work with the first-tier supply chain managers in providing environmental monitoring data access and training (Tachizawa & Yew Wong, 2014). Then, it is up to the individual firm's corporate social responsibility (CSR) policy specifics to engage in training and then oversee compliance with monitoring mandates in the lower-tier suppliers (Andersen & Skjoett-Larsen, 2009).

The main issue is that environmental monitoring is a function of the public sector, not the private sector. Environmental policy is created by governments, specified according to local laws and priorities, and environmental damage is defined in specific metrics of pollution. The film *The World According to H&M* shows exactly how this reality of policy independence creates the loopholes that can provide retailers such as *H&M* with that opportunity to claim false sustainability credit, the very definition of "greenwashing" (Maurice & Hermann, 2017).

Filmed a year after the tragic collapse of Rana Plaza complex,[1] the documentary explores the grounds around another facility in Bangladesh, owned by the *Shinest Group* – an approved supplier for *H&M*. The camera tracks liquid effluents in solid blues, purples and pinks, draining through massive pipes away from the factory into a school yard. The school yard is soaked in colored pools of dye and children are shown splashing in them while playing soccer. The yard is also half full of discarded textile waste, making it look more like a landfill than a school yard. These few minutes of film capture the essence of the reality that remains unchallenged. The reason is that the entity that must challenge it is the local government. It must provide the policies of environmental monitoring, zoning, health and occupational standards, and enforce them.

A school yard is the epitome of a social vulnerability. If the government allows toxic sludge to be pumped directly there, no external body – not *H&M*, not another government, nor an army of corporate social responsibility managers – can infringe on its sovereignty and make demands. What such bodies can do is not engage in further business dealings there. The film explains why that is not happening. The reason is simple – these few shots in that documentary are the only evidence of the dynamic – and minutes into the filming, the camera shows armed guards swooping in on the crew, making them turn of the cameras. Later in the film, when posing questions about their observations to the corporate spokesperson for *H&M*, the answer is that the company is not aware of every-day operations in their sourcing markets, including Bangladesh, and that oversight is under the scope of local authorities. The only thing *H&M* can do is ask for compliance, which Helena Helmersson – Chief Sustainability Officer – assures the investigative team that they do.

None of it seems to be true either in Bangladesh or other *H&M* supplier production clusters shown in Ethiopia and distribution centers in Poland. Furthermore, overwhelming evidence is presented of forged documents and false paper trails by factory owners and managers. Miss Helmersson is shown to refuse to look at the documents, footage, pictures and testimonies of workers, dismissing the charges with rehearsed vacuous responses.

Helena Helmersson is a master at deflection. In the last several years she has been the face of *H&M* corporate social promotion at the most notable fashion sustainability forums, including the annual Copenhagen Fashion Summit, where she was famously challenged in 2014 by Livia Firth, Creative Director of Eco Age and founder of the Green Carpet Challenge (Brismar, 2014). Helmersson's refractions of the difficult questions, and Firth's relentless and passionately angry accusations, at the event became the news of the summit in industry circles. The exchange is a powerful example of the main challenge the industry is facing – telling the truth about the nature of its sustainability problems, while trying desperately to create and maintain a positive image for the brands.

Misinformation about sustainability has been the strategy for promoting a socially conscious message and an image of care for all retailers. For one thing, misinformation is a function of the ambiguous meaning of the term sustainability. When used in public relations messaging, "sustainability" is presented in gradations with such phrases as "more sustainable practices" "more sustainable operations" "more sustainable future." In the BBC documentary *The Sustainability Challenge*, *H&M's* representative Giorgina Waltier repeats the terms "sustainable" and "sustainability" incessantly while honestly answering, for the first time on record, that the company has not made a notable "jump" in offering products with a lower carbon footprint. She explains that the necessary platform – a technology that would allow cotton to be recycled to scale, as cotton is the most essential fiber in all fabrics – is not there yet (BBC World News, 2019). If that is the case, the journalist asks, "will you" (meaning *H&M*) "cut down on the amount of product you offer?" Waltier replies with:

> "We are not going to cut down on the amount of product we offer. We are going to ensure that the product we do offer is made in the most sustainable way… We do respond to consumer demand and our goal is to produce better clothes… We want everyone to have access to sustainable fashion; to sustainable clothes."

The conversation moves on with that vacuous note, with no clarification of what is "more sustainable" or "sustainable" period, according to *H&M's* standards. The message *H&M* wants to send is clear. It is that the company is committed to sustainability. Still, *H&M*, and the industry at large, very seldom bother to detail exactly what is meant by "sustainability."

The term is brandished tirelessly without context. It is this problem that at the 2019 Copenhagen Fashion Summit, the Union of Concerned Researchers in Fashion specifically details in its launch memorandum: "The Union of Concerned Researchers in Fashion wishes to highlight the paradoxical or even misleading use of language in describing 'sustainable fashion' activity" (UCRF, 2019). The organization, comprised and expanding its membership of academics and professionals from around the world, acknowledges that the irresponsible use of the term "sustainable" has deflected from efforts to incentivize research, development, and innovation in production.

The promotion of "sustainable" initiatives and commitments is a promise to change that so far has gone unfulfilled. It is also an issue that the Norwegian government has called *H&M* out on as recently as August 2019.

Covering the story for *Fortune Magazine*, Dwyer (2019) quotes that the deputy director general at Norway's Consumer Authority in stating:

> "Our opinion is that H&M is not being clear or specific enough in explaining how the clothes in the Conscious collection and their Conscious shop are more 'sustainable' than other products they sell. Since H&M is not giving the consumer precise information about why these clothes are labelled Conscious, we conclude that consumers are being given the impression that these products are more 'sustainable' than they actually are."

The official response from *H&M* to the probe in Norway was that it is pleased to be in dialogue with the Norwegian Consumer Authority "... regarding how we can become even better at communicating the extensive work we do" (Hart, 2019).

Statements of commitment to sustainability are now the official public relations message of the whole industry. The commitment toward transforming existing production practices with the goal of making environmental improvements is exemplified with the launch of the Fashion Industry Charter for Climate Action, under the scope of the United Nations' Climate Change program in 2018 (UNFCCC, 2018). The charter's goal is to "achieve net-zero emissions by 2050," a commitment signed by 43 entities including the largest global apparel brands, consortia of professional associations, lobbying bodies, and logistics firms. The charter outlines 16 "principles" and "targets" that can be broadly summarized into goals to change business operations on two fronts: 1) "decarbonization," as is the term, of production and logistics, meaning using cleaner and carbon-neutral materials and shortening the long transportation distribution chains of the industry and 2) to use the language of the charter "improve consumer dialogue and awareness." This specific goal is outlined to target two issues. The first is improve clarity in communicating "sustainable" platforms in order to generate customer support for change. The second is to look for support in policies that would seek to finance the needed changes. As the charter puts it, "scalable solutions" are needed. As the next chapters aim to clarify, "scalable" is "expensive" and therefore hard to finance at current industrial profit margins.

The main goal of the UN Fashion Industry Charter on Climate Action is to create a system for the development of alternative to the current business models. As already explained, the three main fashion business models are luxury retail, branded retail, and fast fashion retail. The charter vies to look for ways to incentivize the creation of a fourth business model to replace all three – the "circular" fashion business model. As a concept, it is worth explaining the evolution from "sustainable" to "circular." There are major differences that help elucidate the difference between commitment toward change and actually making changes in production and operations.

Promoting Circularity

As "sustainable fashion" has become a phrase, it is so pervasively used that one would be hard pressed to find the term "fashion" not accompanied by an adjective denoting a sustainability component. Terms such as "eco" fashion, "green couture," "circular fashion," "sustainable design," converge in parlance and media propensity under the newly coined platform of "circularity." Circularity is… What is it? It is often used to describe a business model. Linder and Williander (2017) analyze the evolution of the term and its acronym CBM – circular business model – in juxtaposition to the traditional production management operations model known as EOM – original equipment manufacturer production.

The core difference is in the concept to "reuse" already manufactured inputs. The authors track early works on CBM deployment from about a decade ago that provide hopeful estimates that, across industries, as much as 80 % in material and energy use reduction can be achieved by adopting circular policies. Then the authors show why such hope has not created an economic revolution of circular business model explosion.

The main reason is the difficulty to "reuse" as "recycled" inputs have serious property limitations. They are fragile, have lower durability, and are of poor quality. Therefore, most products manufactured with recycled components have a very low percent of actual recycled material in them (Geyer, Jambeck, & Law, 2017). In apparel, the reality is that less than 1% of all product sold comes from any form or recycled or reclaimed input (Rathinamoorthy, 2018).

Be it plastics, paper pulp, or fabric, all recycling is done in expensive petrochemical processes that are themselves polluting (Gupta & Gupta, 2019; He et al., 2015; Pensupa et al., 2017). However, from sustainable fashion promotion language one would never know it. The industry has launched a wave of recycling-promoting initiatives. Once again, *H&M* is in the front of visibility, with the launch of its "Conscious" collection, signing as its face, eco-fashion activist Hollywood actress Olivia Wilde in 2015. The same year *H&M* launched its global recycling initiative with the promotional "film," is the term the company used, but it is more of an infomercial, "There are No Rules in Fashion but One: Recycle Your Clothes." And in 2018, *H&M* expanded the "Conscious" collection with "Conscious Exclusive," promoting more product lines made "sustainably."

Young (2018) describes the "sustainable" innovations to explain that, alongside previous "Conscious" product lines made with recycled wool and organic cotton and linen, there are clothes also made with Tencel, which is the leading branded lyocell – a material that has evolved from the viscose rayon process using fibers spun generally from beech or eucalyptus. *H&M* claims its Tencel is produced in a "closed loop" system that reuses "nearly" all the chemical solvent it requires. This is a very dubious claim. Once used, chemical solvents – as the term suggests – are "dissolved." They cannot be filtered back from their dissolved state and re-used. One cannot simply reuse poured bleach.

The "Conscious Exclusive" collection also offers garments manufactured with recycled polyester and nylon. These are positive and important steps forward in product development, but arguably not enough to brand the entire product line "sustainable." The issue is that the items sold are made "with" those inputs, not exclusively from them. Additionally, it does not address the myriad of treatment steps in integrated production, other than the raw material for the main fabric. A Tencel or organic cotton t-shirt, colored with a heavy-metal laden dye, printed with a plastisol ink, should not be able to fall into this category. But, according to *H&M's* standards of "sustainability," it appears that it could.

The "Conscious Exclusive" collection is promoted masterfully with the face of the 1990s supermodel that has become a cultural icon in the industry because of social justice, humanitarian and public health activism – Christy Turlington. As an activist, Turlington is best-known for her humanitarian work on prenatal care for disenfranchised women and her anti-smoking campaigns. She has launched her own product lines in wellness merchandise, promoting public health activism. Turlington has a strong, even unquestionable "eco-brand," as is the term used to describe celebrity promotion power. She is also of a generation that represents the mothers of today's *H&M* core demographic.

The launch of the collection is a public relations masterpiece. It showcases not just a commitment to sustainability, but an intergenerational innovation in fashion branding. It uses a celebrity spokesperson outside of the age bracket of its main audience, clearly targeting a different market – those older and more socially conscious consumers that have been fleeing fast fashion stores. The different visual and advertising copy incarnations of the campaign shows that one message dominates above all else – recycle.

The promotion of the concept of recycling is so popular, few question it. It also calls to mind the first *Levi's* lifecycle assessment of a pair a jeans, which concluded that consumer care is the most ecologically impactful stage (Westervelt, 2012). Both messages aim to shift responsibility for the environmental impact of fashion to the consumer. The truth is that with current technologies, it is nearly impossible to fully recycle clothes in a way that can constitute an ecological improvement. Mechanical recycling, an option for natural fibers, will often result in weaker fibers that must subsequently be blended with unrecycled fibers. Chemical recycling, as the name implies, is a chemically as well as energy intensive process. It can be used to recycle synthetic fibers such as polyester.

Synthetic fabric can be both chemically and mechanically recycled (Lv et al., 2015). Mechanical recycling takes less energy, but results in poorer quality. Chemical recycling breaks the fibers down into the original components for repolymerization, and in theory, results in a quality that is a good as virgin fiber. However, challenges still exist. A major one is separating out blended fibers for different recycling processes, and another is color sorting (Zou, Reddy, & Yang, 2011).

Only as of very recently, as is to be explained in the *Re:newcell* example in Chapter 6, has chemical recycling been explored and scaled for fabric made from natural fibers in a process that captures the cellulose which can then be used to create a man-made cellulosic fiber - essentially a viscose. Viscose is the technical term used to describe variety of fibers made from the regenerated cellulose that is left after the chemical decomposition of natural fibers.

The recycling of natural fiber yarns involves stripping dyes with solvents that produces toxic, and untreatable through filtration, water discharge (Gordon & Hsieh, 2006). The least toxic option available at this point is in the recycling of jeans where stripping can be replaced by a mechanical process, but only if the jeans were died with indigo, which is a consistent dyestuff (Paul, 2015). For all other fabric made from natural fibers, after stripping, heat-intensive treatment is needed to create pulp from which new fibers could be spun. Those fibers then need to be strengthened with alkali and starch to impart stiffness, which are also organic polluters expelled into waterways (Wanassi, Azzouz, & Hassen, 2016). The resulting product is fragile yarns that must be rewoven with unrecycled stronger yarn in order to be durable, versatile, and able to withstand the follow-up dyeing and finishing processes. Conducting on-site evaluation of the properties of such yarns, Halimi, Hassen & Sakli (2008: 787) frankly state:

> "There are many published papers that have discussed the cleaning behaviour of virgin cotton and its characteristics. Never the less, there are only a few published papers about recovered fibres quality and there are no published studies about cleaning behaviour of cotton waste. These kinds of studies are crucial to encourage manufacturers to treat cotton waste."

This statement, made over a decade ago, about the state of blended/recycled natural fiber research is still valid. Textile mechanics research on recycled/blended yarns is absent. There is little evidence that manufacturers have incentives to treat cotton waste, as Halimi, Hassen, and Sakli (2008) admonish. Yet, retailers have embraced the recycling rhetoric, misleading the consumer into believing that it is commonplace to recycle clothes into clothes, when in fact this process is only just beginning to happen on a pilot or exhibition scale.

Note

1 Although the credit citation of the film in the list of references is 2017, that is the year the film was readied for international release. It first aired in France in 2014.

Chapter 4

The Realities on the Ground

Fashion's Ecological Problem

Since the introduction of the fast fashion business model of retailing, the apparel industry has grown and continues to grow significantly. Estimated at between 2.4 and 3 trillion dollars (Amed et al., 2017; BBC, 2018), it is expected to grow by an additional 63% by 2030 (Kerr & Landry, 2017). This growth has raised sustainability concerns. Waste from the frequent discarding of used clothes is identified as the main problem (Cline, 2012). Retailers have responded by offering solutions to reduce waste by promoting and even offering recycling and upcycling services (Birtwistle & Moore, 2007; Joung, 2014). However, as already noted, it is technologically nearly impossible to recycle clothes in a way that can constitute an ecological improvement. Fabric recycling is a chemically and energy intensive process that can be used for the treatment of fabric made from natural fibers, mainly cotton and wool. Synthetic fabric could be melted back into basic plastics (Lv et al., 2015), yet there is no evidence of commercial operations that do so at scale. Blended fabrics – those made from the combination of natural and man-made yarns – cannot be recycled (Zou, Reddy, & Yang, 2011). Most importantly – recycling is expensive.

In mass production, "new-growth," as is the term, yarns are the main input of all fabrics. New-growth yarn is mainly made from fresh – "new" – unrecycled cotton. The reality is that the technological platform for the production of cotton has largely remained unchanged for a century. It is reliant on natural resources, mainly agricultural land and water for growing cotton, water for its manufacturing, and water for its care and use. From the initial stages of planting cotton seed, to the retail stage, it takes over 20,000 liters of water, which is 5,800 gallons, to make 1 kilogram, or 2.4 lbs. of cotton (BBC, 2019). That is enough for one t-shirt and one pair of jeans.

Cotton is essential because it is the cheapest natural fiber and it blends well with synthetic fibers (Minot & Daniels, 2005). It is the backbone of all yarns because of its porous structure that allows for the creation of blended yarns, or "poly-blends," or simply referred to by many industry insiders as "blends." Cotton and synthetic blends have become the most-used apparel material (Ramamoorthy, Persson, & Skrifvars, 2014). It is the combination of cotton and ethylene terephthalate (PET) yarns that drives poly-blend innovations to improve durability, versatility, and finery in apparel (Chen & Zhao, 2016; Hussain et al., 2016; Koh, 2005).

Research has shown that consumers do prefer clothes made from and with cotton (Birnbaum, 2005; MacDonald et al., 2010), and with the growth of promoting sustainability, cotton is benefiting from some "greenwashing" PR. The "Fabric of Our Lives" social media platform of the US Cotton Council boasts environmental stewardship with the statement that cotton is a "natural choice." It is, yet its agrarian production is much criticized for high fertilizer toxicity (Dem, Cobb, & Mullins, 2007; Oosterhuis & Weir, 2010), giving rise to the promotion of organic cotton.

Organic or inorganic, cotton production accounts for around 6% of final apparel costs (Ferrigno et al., 2005). Weaving cotton into fabric is more than twice as expensive because it is an integrated and technology-dependent petrochemical process. Weaving brings the costs of textile production, including farm-level costs, to account for 13% of apparel item retail prices (Birnbaum, 2008). The weaving of any fabric – natural, man-made, and blended – is also environmentally taxing because of the pollution from the necessary wet finishing processes.

As noted in Chapter 3, the main wet processes are scouring and bleaching, desizing, mercerizing, dyeing, garment finishing, laundry, printing, and coating (Wakelyn, 2006). The combination of the wet processes, also called textile "finishing" processes, happens at large volumes in textile mills that use river streams for their water needs (Angelis-Dimakis, Alexandratou, & Balzarini, 2016). Because mostly such waste cannot be treated at scale, it is not surprising that empirical evaluations have shown that mills around the world do not have adequate arrangements to treat run-offs. While the WTO has proposed standards for tolerable discharge levels of effluents, its water quality standards are classified into aggregate measures and maximum allowable concentrations of specific chemicals in point-of-discharge run-off drainage. The problem is that the standards for these measures, well established in effluent regulation and well researched from across industrial sectors (Banuri, 1998; Bhutiani, Varun, & Khushi, 2018; Hessel et al., 2007; Molla & Khan, 2018; Prasad et al., 2015), are mere guidelines.

Among the direct costs associated with effluent discharge compliance is the nebulous determination of accurate measures of water pollution. The WTO breaks down the aggregate measures into specific categories based on available test capability. They are pH value, which determines acidity or alkalinity in run-offs, temperature, biological oxygen demand (BOD), chemical oxygen demand (COD), total suspended solids (TSS) or non-filterable residue, total dissolved solids (TDS), and color (Banuri, 1998). The problem is that, unless specific testing is asked for and administered by local government, the collecting and reporting of these pollution measures is simplified in regulation and oversight. The World Bank only reports BOD levels – the most general pollution estimation. The reason is the ease and specificity of the BOD test itself – it is simple.

BOD values offer an approximate measure of water "cloudiness" which tells the tester very little of what causes the cloudiness and in what amounts. The other tests require more complicated laboratory operations because TSS must test for each individual pollutant by itself. As noted in Chapter 3, the main pollutants are Mercury, Lead, Manganese, and Cadmium and the heavy metals Zinc, Copper, and Iron (Molla & Khan, 2018). Of those, the heavy metals are TSS, which can be filtered out technically (at least in laboratory settings). Total dissolvable solids on the other hand are classified as organic and inorganic pollutants that are left in water even after it is filtered. In fashion run-offs they include the highly toxic Mercury and Chlorine. Total dissolvable solids can be run-offs from a wide variety of sources, not just a given textile mill, including agricultural use, mining, and even road salting (Chapman, Bailey & Canaria, 2000).

As biochemical research on these elements' concentration in run-offs from textile manufacturing establishes both them as source pollutants and their tolerable concentration levels (Howard, 1933; Goodfellow et al., 2000), the testing process happens in scientific labs. From there, the information must be put in regulatory practice by local governments through their environmental bureaucracies. In actuality, that would mean having to conduct daily tests for each of those pollutants around each production facility, most ideally several times a day, in order to gauge compliance. When infraction is noted, the overseeing body issues a citation, then the government fines the facility. That is how the system should work but it does not.

The main reason – it is expensive – and in apparel, costs are to be kept low above all else.

As research mentioned so far has shown, environmental regulation in textile production is inadequate. As the industry grows, heavily reliant on the fast fashion retail model, the escalating demand for cheap fabric magnifies price-cutting pressures on textile production (Adhikari & Yamamoto, 2007). When the advent of fast fashion quintupled the demand for textiles (Amed et al., 2017; Anguelov, 2015; Caro & Martínez-de-Albéniz, 2015; Jeacle, 2015), how have producers coped with the lack of technological options to meet effluent toxicity guidelines while increasing production? Business precedent would suggest that an alternative would be to circumvent guidelines through pollution havens. The term "pollution haven" explains the relocation of industrial facilities to evade environmental regulatory stringency.

Dealing with the Toxicity of Textile Manufacturing

The "pollution haven" hypothesis states that a large proportion of foreign direct investment (FDI) in lesser-developed countries (LDCs) finances highly polluting and ecologically inefficient manufacturing processes and facilities that are outsourced from developed countries (Grimes & Kentor, 2003; Jorgenson, 2007; Lee, 2009). The pollution haven hypothesis is applied to fashion because most textile production is done by multinational corporations (MNCs) from developed nations that manufacture in developing nations (Miroux & Sauvant, 2005). Therefore, FDI is the backbone of the industry. But why? Lower labor costs are the most often-cited reason (Mihm, 2010; Taplin, 2014; Tokatli & Kızılgün, 2009). Yet, modern textile production facilities do not employ that many people.

Malik et al. (2010) analyze working conditions in textile mills in Pakistan and offer a sample of mills (see Table 4.1) where total worker numbers range from 220 to 750. "Large-scale" mills are typically defined as employing around 1000 people (Bruce & Daly, 2004; Chavalitsakulchai et al., 1989). One must discern the drop in numbers from research made two decades ago to more recent works where the number of workers in mills are notably lower. It is due to mechanization and robotization. Fewer and fewer workers are needed to make fabric. Where the numbers remain steady is in assembly.

Even in assembly there is a worry about job loss from robotization, as innovation is noted with the invention of "sewbots" (Graham-Rowe, 2011;

Guizzo, 2018). The technology is not at scale yet. If it develops to that level, it would be based on significant financial investment from apparel MNCs.

Miroux and Sauvant (2005) assert that MNCs dominate global production and in the developing world their affiliates dominate the sector. As a result, developing countries have accounted for a rising share of the growth in textile and apparel exports so much so that by 2005 they produced half of all global textile exports and nearly three-quarters of global apparel exports (Andriamananajara, Dean, & Springer, 2004; Miroux & Sauvant, 2005). That is the case because of the large volumes in production of large-chain, global retailers.

Large retailers have large volume requirements. Thus, they only consider large suppliers, which leads to the increasing role of MNC foreign investment, as conglomerates are looking to expand capacity. Growth of capacity is dependent on the ability to attract capital. Producers in developing nations have limited financial and know-how capabilities, therefore, an increase has been reported in the foreign ownership of both textile mills and garment manufacturing facilities, particularly those that employ more than 1000 workers (Bruce & Daly, 2006).

For certain countries it is "textiles" and not "apparels" that defines exports. The export classification "textiles" refers to the making of fabric, while "apparels" refers to the assembly of finished garments. For an in depth discussion of the difference between the sectors, see Kunz, Karpova and Garner (2016). For example, in Pakistan, one of the leading exporters of both textiles and apparels, textiles have grown to comprise over half of all merchandise exports. In India apparel exports account for 55% of all export earnings. However, only about 12% of those exports are in the form of ready-made garments so that 88% of exports classified under "apparel" are actually in the form of fabric (Chaturvedi & Nagpal, 2003). The other global leaders in textile exports are Nepal (16%), Macao (China) 12%, Turkey (11%), and India (11%) (Miroux & Sauvant, 2005: 4).

Fifteen developing nations including China, India, Pakistan, Bangladesh, Egypt, and Turkey account for 90% of global textile exports and 80% of global clothing exports (Adhikari & Yamamoto, 2007). Among them China has risen as the leader in the industry and is referred to as "the tailor of the world" (Mikic, Adhikari, & Yamamoto, 2008; Pan et al., 2008). Chinese textile companies are the largest in the world, but still over 34% of Chinese textile and apparel exports come from Chinese enterprises financed by foreign investors (Miroux & Sauvant, 2005). In Indonesia, over 90 % of textile mills are owned by foreign investors (James, Ray, & Minor, 2002;

Robinson, 2008). The *Toxic Threads* reports from Greenpeace show that those mills in Indonesia emit pollution with little regard for the environment (Greenpeace, 2012).

The fact that textile mills in the developing world are at large under foreign ownership has raised ethical and environmental concerns that textile MNCs are strategically locating in countries that are still developing environmental regulatory systems in order to exploit regulatory uncertainty (Greer et al., 2010; Khan et al., 2009). It is important to understand what this "regulatory uncertainty" entails. In textile production in the developing world, it is a lack of effective local environmental regulatory system. When MNCs from well-developed nations choose where to locate, it would be naïve to not question whether they would face incentives to locate to nations where they would be allowed to pollute without limits.

As society, we have known for decades that this process is at play and that we all reap economic benefits from it. It allows us, in the developed world, to enjoy cleaner environments while paying lower prices for our goods. It was sociologists (Grimes & Kentor, 2003; Jorgenson, 2007, 2009; Rice, 2007), following on the pioneering work of Chase-Dunn (1975) that started the academic analysis of empirical evidence of industrial pollution from FDI. The field has pioneered the concepts of ecostructural investment dependence, arguing that a large proportion of foreign investment in LDCs finances highly polluting and ecologically inefficient manufacturing processes and facilities that are outsourced from developed countries. Across academic disciplines that charge is referred to as "the race to the bottom phenomenon," the "theory of ecologically unequal exchange," and, as previously noted, the "pollution haven hypothesis" (Frey, Gellert, & Dahms, 2018; Gray, 2006, Jorgenson, 2006, 2007, 2009, 2012; Kentor & Grimes, 2006; Shandra, Shor, & London, 2008; Smarzynska & Wei, 2001; Wagner & Timmins, 2009). It is an overall theory of pollution displacement that has increasingly generated interest since the early 1990s, yet Elliot and Shimamoto (2008) argue that earlier studies have found little empirical support. Supporting evidence comes from fairly recent cross-national studies of greenhouse gas (GHG) emissions and other forms of air and water pollution.

Theory vs. Practice

In terms of empirical findings, the seminal works on pollution havens come from environmental economics with models built with data mainly from the developed world (Brunel, 2017; Hille, 2018; Lee, 2009; Levinson & Taylor, 2008; Millimet & Roy, 2016; Wagner & Timmins, 2009). Brunel and Levinson (2016) offer a thorough review of extant studies, which are almost exclusively from OECD nations – Organization for Economic Co-operation and Development. The general argument for using these relatively well-developed countries as units of analysis is data availability with a particular focus on metrics that can be used as proxies for environmental stringency. Two factors are common in these studies. They are: (1) a general assumption that environmental regulation is a robust and developing social and political force, as covered by the works on Environmental Kuznets Curves (EKCs) (Al-Mulali et al., 2015; Dinda, 2004, Shahbaz et al., 2015a) and (2) stringency can be followed by producers, albeit at increasing fixed costs. These assumptions are not transferable to the developing world, and especially LDCs with developing institutions, where evidence has shown that environmental regulation is not a social and political priority (Cole & Elliott, 2003; Pattanayak, Wunder, & Ferraro, 2010). The assumptions are also not transferable to the apparel sector because, as explained in the previous section, environmental stringency is ill-defined and lacks binding regulatory features, mainly because it cannot be followed by producers – textile pollution cannot be treated.

In terms of examining pollution havens in developing nations, research with assumption transferability to the apparel sector comes primarily from sociology. Sociologists refer to the interplay between displacing pollution from the developed to the *industrializing* developing world as the theory of "ecologically unequal exchange" rather than "the pollution haven hypothesis" (Jorgenson 2007, 2012; Moran et al., 2013; Rice, 2007; Roberts & Parks, 2009; Schofer & Hironaka, 2005) although some of the works note the link and use the terms interchangeably.

Such works show that in the past several decades rising pollution is associated with the growth of foreign direct investments in emerging markets across industries (Acharyya, 2009; Jorgenson, 2006, 2007; Lucas, Wheeler, & Hettige, 1992). Environmental issues are especially evident in developing nations reliant on structural adjustment loan programs from the World Bank and IMF (Chaturvedi & Nagpal, 2003; Easterly, 2005; Reed, 2013). That is the case because they compete more aggressively for FDI in order to make structural adjustment loan payments (Grimes & Kentor, 2003; Jorgenson, 2007, 2016; Kentor & Grimes, 2006; Lee, 2009). An industrial policy in those nations toward increasing exports especially increases pollution (Cherniwchan, Copeland, & Taylor, 2017; Stretesky & Lynch, 2009).

The commonality in these studies is the focus on correlation of environmental damage and foreign capital penetration in the general absence of local proxies for environmental regulation that are comparable in detail, magnitude, or enforceability to those in the developed world. Examining panel data and time frames before the current academic trend to create instrumental variables to capture environmental stringency, the filed notes significant problems with endogeneity and heterogeneity. Endogeneity here stems from difficulty to separate cause and effect. It is not easy to see if investment causes pollution to increase or if polluters choose to locations that allow them to pollute. Heterogeneity is an issue when policy and economic dynamics in one nation, for example, are very different from those in another.

In both bodies of literature – Western-centered "pollution haven hypothesis" and LDC-focused "ecologically unequal exchange" – the empirical proxies for environmental damage are most often air pollutants. In the pollution haven literature, even more recent works such as that of Cherniwchan, Copeland and Taylor (2017), which provide a review of published articles with examples from the developed and developing world, are still exclusively focused on air pollution emissions as proxies. The explanations of proxy choices are data availability and the priority of government regulation toward reducing GHG emissions. That is the reason why governments collect and report air pollution data more stringently than other toxicity metrics.

An important contribution of Cherniwchan, Copeland and Taylor (2017)'s work is the discussion of implications of previous findings in making the distinction between the pollution haven hypothesis and the pollution offshoring hypothesis. This distinction is important in the context of the apparel sector because it clarifies that "domestic" firms can become less polluting when out-contracting the more toxic production processes to foreign partners. In apparel, that means purchasing rather than making yarns and fabrics. An additional clarification on that article's analysis must be noted here with respect to the applicability of its theoretical assumptions to the apparel sector. The issue is with the assumption accuracy of the specification of technology and cost of implementing environmental mitigation. In equation 20, $z = g(A)D$, where z stands for pollution and D for dirty cheapest input option, the authors posit that z can decrease as a function of firms paying a fixed cost (A) toward investing in abatement technology (Cherniwchan, Copeland, & Taylor; 2017: 72). As already explained, that is not possible in apparel production. Textile producers cannot invest toward abatement technology because there is no abatement technology (Kant, 2012). They can invest toward developing cleaner products, but as of yet, lack economic incentives to do so. The issue of integrated ecological damage from fashion commerce has just of late begun to gain social and academic attention.

Not until recently were textile effluent discharge levels even included in major pollution counts by the academy. For example, covering data from 1960 to 1995, Mani and Wheeler (1998) rank the 10 most water-polluting industries. In that and other such rankings (Oketola & Osibanjo, 2007) textiles is either omitted, or included, in "miscellaneous manufacturing." The stark difference is evident in the fact that back in the 1990s textile pollution is not mentioned in environmental research. Today studies focus on it with claims that it is a leading global organic water polluter (Desore & Narula, 2018; Hasanbeigi & Price, 2015; Nimkar, 2018).

The aforementioned literature may provide some understanding as to why water toxicity from textile production is not well covered by the academic works from environmental economics. First, it was not on social and academic radars until the industry started to grow at unexpected rates in the late 1990s. The fast fashion model quintupled production in a span of little more than a decade and its rates of growth keep increasing (Caro & Martínez-de-Albéniz, 2015). Another theme emerges in the theoretical assumptions of pollution haven research from environmental economics – that of a hypothetical presence of an autarky – a nation that may choose not to trade. In apparel that has not been an option since the implementation of the Multi-fiber Agreement (MFA) in 1974. It established a system of trade in components and partially assembled apparel for diplomatic, rather than economic reasons (Anguelov, 2015). The MFA dictated choice in location based on quotas each exporting country was given for exports to the main Western markets.

An additional generalizability issue in previous pollution haven and ecologically unequal exchange works arises from the fact that they use data mainly from three industrial sectors – energy, metallurgy, and manufacturing. These are sectors well known for innovation in environmental mitigation technologies. In such a context, the assumption that lowering environmental damage is an option through deployment of technology (albeit at increasing fixed costs levied on producers through policy stringency mandates), could have merit. However, more recent works on environmental stringency show that even in such cases, transaction costs of technological deployment of pollution mitigation may not make it feasible to reach stringency policy objectives. Specifically, the literature on the stringency of Renewable Portfolio Standards (RPS) shows the challenges when incentivizing the generation of renewable energy by government (Carley et al., 2017, 2018; Upton & Snyder, 2015, 2017) meets with the reality that there is no commercially viable technology (yet) to store it at scale (Anguelov & Dooley, 2019). The outcome is a relatively slow divestment trend from fossil fuels in energy generation despite significant growth in the diffusion of the type and scope of regulatory policy innovations toward increasing environmental stringency mandates. It is due to social definitions and economic priorities in the short- and long-term goals of addressing environmental damage.

Relaying the complexities in quantifying environmental damage, Brunel and Levinson (2016: 46) discuss the "multidimensionality" of the definition of "environment." Chapter 5 of this book is to offer an analysis of such complexities, by tracking the evolution of the works studying EKC. The evidence shows that environmental stringency effectiveness continues to be problematic precisely because environmental priorities are value-laden. Different individuals, governments, and nations value them differently.

As a way forward in research strategies of this complicated interplay, one study stands out. Keller and Levinson (2002) posit that choosing US states as units of analysis is a good strategy because of access to unit-comparative data on foreign assets and environmental stringency. In academic logic, this strategy makes sense on two fronts. One is indeed being able to tackle heterogeneity, as American states, despite their differences, are still a politically cohesive unit of analysis. The other is that the United States has historically been the source of environmental policy innovation diffusion, meaning laws implemented in America can be and are often copied in other nations. However, such a hopeful logic, and even examples of this dynamic at play, does not mean it is the norm, as Chapters 5 and 6 of this book show.

Keller and Levinson (2002) use abatement costs as a proxy for environmental stringency and foreign-owned plants (with a discussion on the difference between greenfield and merger-and-acquisition FDI) as the main dependent variable. Abatement costs are the combination of purification technology investments by firms, as mandated by (in this case) state governments, and/or fines imposed on the firms for environmental damages, such as toxic spills or accidents. Yet, it becomes unclear through the analysis how FDI can impact the leading pollution sector identified by the authors – pulp and paper. Therefore, the logic of the study falls short, as overall state-level abatement costs, if they are indeed a function of the leading pollution economic sector – pulp and paper – may not be borne by many, if any, foreign firms.

FDI in America has mainly been directed toward knowledge-intensive sectors, such as finance and professional services, or market-access manufacturing subsectors, such as automotive, engineering, and information and telecommunication technologies, to follow the Bureau of Economic Analysis (BEA) industrial classifications for that time (Chung & Alcácer, 2002; Keller & Yeaple, 2009; Love, 2003). Little, if anything, has ever been said about foreign firms acquiring or building paper mills in America. Furthermore, since then, the pulp and paper industry has been leaving American shores (Nagubadi & Zhang, 2008; Zhang, 1997). Specifically, Uusivuori & Laaksonen-Craig (2001) find that in the 1990s US outbound FDI in "forest" industries, which includes pulp and paper, reached substitution levels with exports.

The reasons are multidimensional, to use the term of Brunel and Levinson (2016), but the main environmental factors behind the exodus are also behind the decline of American textile production (Leiter, Schulman, & Zingraff, 1991; Minchin, 2012; Moore & Ausley, 2004). They are increasing stringency with the passing and augmenting of the Clean Water Act (Houck, 2002; US EPA, 2019), regulating on the volume of effluents (Karr & Yoder, 2004; Wenig, 1998) rather than their content, and most importantly, lack of purification options (Kant, 2012). Without efficient filtration technologies, it is not feasible for either the pulp and paper or textile sectors – reliant the ability to expel bleaches, solvents, acids, and dyes in large quantities into waterways – to operate at global market-competitive volumes, while adhering to trace-element emission directives in the United States (Luken, Johnson, & Kibler, 1992; Norberg-Bohm & Rossi, 1998). Based on these dynamics, the logical research question arises: "Are textile MNCs choosing pollution havens?"

This question must be predicated with a condition that location choice is an option. For decades, for American textile firms in particular, it was not. From 1974 to 2004 restrictive "rules of origin" directives of the MFA for the international trade of apparels and textiles limited the amount of product that could be imported into the United States from specific countries. American (or other textile MNCs that relied on exports to the US market) would have an incentive to locate into nations with available quotas first and foremost. However, with the growth of globalization and the change of the fashion industry from national to one of global brand expansion, producers found the MFA a cumbersome barrier to internationalization. Therefore, under lobbying advocacy, it was agreed to dismantle the MFA gradually by 2005 (Appelbaum, 2005; Appelbaum, Bonacich, & Quan, 2005; Martin, 2007).

The removal of the MFA, in effect, enabled producers to source freely and locate new facilities in sites that offer the lowest direct and indirect costs. Such indirect costs include lax environmental laws and/or the willingness of local governments to not adhere to them. Taking this policy change into account, the question becomes: "Given the ability to choose "pollution havens," are textile MNCs doing so?" Providing an empirically backed answer to this question is not easy because of data availability. Yet, from of the literature examined so far that points to the formation of textile production clusters in particular developing nations, where evidence shows textile factories do not follow regulatory guidelines, it is worth offering an analysis of what little data there is and a discussion as to why data are lacking.

The Data Problem

Indicators of pollution from textile manufacturing were collected at the country level by the World Bank until 2008. Their values are available in the *World Development Indicators* database (World Bank, 2019). The data collection has been discontinued and no explanation from the World Bank is given as to why. The last year with available data is 2008. Yet, with this very limited data, it is possible to conduct a very basic analysis. Values are available as far back as 1991; therefore, the models here examine the time frame 1991–2008. Estimations of the United Nations Conference on Trade and Development (UNCTAD), published in an edited volume by Miroux and Sauvant (2005), posit that since the beginning of the 1990s over 90% of textile production has become concentrated in just 32 nations, including some of the world's poorest. Following up on that claim, the sample here is those nations ranked by their reliance on textile exports.

The methodology explanation from UNCTAD for the ranking is that national account data are used in estimating national economic reliance on textile exports. It is calculated as percent of textile exports in relation to all other exports. Unfortunately, for 12 of the top textile exporters, data are not available. Much of the data reported to the World Bank is voluntarily provided by local government agencies. Some countries limit what information they provide and when. Table 4.1 presents the 20 nations for which data

Table 4.1 List of Data-Reporting Nations with FDI and Textile Sector Pollution Averages. Countries Included in the Model

Country	*Region*	*Net FDI in Millions $US*		*Percent H_2O Pollution from Textile Industry*	
		1991	*2008*	*1991*	*2008*
Albania	Europe	20.00	843.32	59.80	60.19
Bangladesh	Asia and Oceania	1.30	973.49	77.11	77.11
Bulgaria	Europe	55.90	8472.89	20.68	28.04
Cambodia	South-East Asia	33.00	794.09	6.83	59.35
Czech Republic	Europe	564.25	8966.97	15.21	7.40
Dominican Republic	Latin American and the Caribbean	145.00	2884.00	73.07	73.07
Egypt, Arab Rep.	Africa	191.00	7574.00	31.11	31.11
Estonia	Europe	80.00	875.99	23.62	8.78
Fiji	Asia and Oceania	11.13	332.37	38.56	38.56
Indonesia	South-East Asia	1482.00	3418.71	31.61	31.61
Korea, Rep.	Asia and Oceania	–308.00	–0594.00	24.99	9.34
Latvia	Europe	272.00	1092.00	19.93	12.61
Lesotho	Africa	273.23	218.45	90.14	90.75
Lithuania	Europe	30.84	1383.16	23.30	19.33
Madagascar	Africa	13.65	85.00	59.93	58.95
Nepal	Asia and Oceania	19.10	0.93	38.66	38.66
Sri Lanka	South-East Asia	43.32	690.00	43.56	43.56
Turkey	Asia and Oceania	783.00	15414.00	30.27	35.66

are available, with the dependent variable and main explanatory variable. Appendix: List of All Nations, included for comparative purposes, presents all 32 nations ranked by Miroux and Sauvant (2005) with the raw data from the World Bank.

The data show that nations most-often identified in the previously covered literature as the world's largest apparel producers – China, India, Pakistan – and the apvparel assembly powers – Hong Kong and Macao – do not report pollution metrics. It is also unclear from the UN report what percent of total global production happens in each of these nations. This is a serious limitation in extant data collection and future research needs to address it. Previously examined studies with the highest similarity to the production and regulatory dynamics of the apparel sector establish FDI as the proxy for attracting foreign firms based on environmental permissiveness (Jorgenson, 2016). Therefore, the dependent variable here is FDI net inflows. FDI net inflows are defined by the World Bank as "the value of inward direct investment made by non-resident investors in the reporting economy."[1] FDI net inflows show annual fluctuation and, when employed in longitudinal models, can capture trends of increasing (and/or decreasing) attractiveness over time. The problem is that they are not disaggregated by sector in the *World Bank Development Indicators* catalog, which is another major limitation. However, for economies most reliant on one overall sector, it has been shown that FDI is directed to that sector, as it defines national economic comparative advantage (Dowlinga and Cheang, 2000; Ramondo, 2009; Rivera-Batiz & Rivera-Batiz, 1990). In textiles, that dynamic is particularly strong, exemplified by studies in the general literature on industrial upgrading that employ the "flying geese" model (Kojima, 2000; Korhonen, 1994). The textile sector is dubbed the "lead goose," attracting most FDI (Akamatsu, 1962; Brautigam, 2008; Schröppel & Mariko, 2003).

Under these assumptions, and with the available data, an explorative panel analysis is presented here. FDI inflows in the nations most dependent on textile production are regressed against proxies of environmental permissiveness and strategic market importance. FDI is recorded in millions of current $US. The independent variables are:

1) $H_20_{TEXTILEit}{-}_1\ H_20_{TEXTILE}H_20_{TEXTILE}$ – Water pollution from the textile industry as percent of all biochemical oxygen demand (BOD) effluents, which captures "industry shares of emissions of water pollutants from manufacturing activities as defined by two-digit divisions of the International Standard Industrial Classification (ISIC) code" (World Bank, 2019).
2) $H_20_{CHEMICALit}{-}_1\ H_20_{CHEMICAL}$ – Water pollution from the chemical industry as percent of all BOD effluents; and therefore, is of particular interest because of the chemically heavy processes involved in textile processing (Adhikari & Yamamoto, 2007; Kant, 2012). This fact must be examined in the context of short lead times admonitions in the textile supply chain literature (Barnes & Lea-Greenwood, 2010, Barnes et al. 2006; Birnbaum, 2005, 2008, 2015; Bruce & Daly, 2004, 2006). If producers do not have time for lengthy shipping and receiving time frames in retail products, would their suppliers feel the same pressures in the sourcing of chemical inputs? If yes, that would mean locales with healthy local chemical sectors would be preferable to investors.
3) $CCD_{it}{-}_1$ – Adjusted savings: carbon dioxide damage (current $US). The World Bank defines Carbon Dioxide Damage as "$20 per ton of carbon times the number of tons of carbon emitted." It is included as an overall measure of environmental permissiveness, as explained in Roberts

and Parks (2009), which estimates the economic damage of pollution. The raw numbers come in estimates to the last dollar. For simplicity here, they are recalculated into millions $US.

4) $AG_{IMPORTS}$ – Agricultural raw materials imports as percent of merchandise imports. It is included to gauge the importance of imported cotton. Birnbaum (2008) estimates that over 80% of exported raw US cotton ends up returning to the US in the form of ready-made garments. Raw cotton is included in the broad category.
5) WTO–WTO membership – a dichotomous variable tracking whether a country is a member of the WTO and also when it joined. Many of the nations in the study joined the WTO at different points during the examined time series. A "0" is assigned for the years in which the country is not a WTO member. A "1" is assigned for the years after which it is a member. Using this coding approach allows us to explore whether FDI increased after a nation joined the WTO.
6) GNI – Gross national income (GNI) per capita. It is included as a personal wealth proxy in the absence of data for wages, under the assumption that in nations with fairly low GNIs wages would be lower. In the general fashion economics literature, the importance of cheap labor for foreign investors is well documented (Anner, 2011; Shaw et al., 2006)
7) $POP_{\$2}$ – Percent of the population living on less than $2 a day. It is included as a national wealth proxy, rather than the often-used GDP measures, under the claim by Miroux and Sauvant (2005) that textile production is concentrating in very impoverished nations.
8) Region – The regions are Continental Europe =1, Asia and Oceania=2, South-East Asia=3, Africa=4, South America and the Caribbean= 5. Regions are entered into the model as dummy variables, with South-East Asia as the holdout group.

South-East Asia is chosen as the holdout region because of the arguments in the literature that fast fashion's short production lead times are necessitating a move of facilities away from that region and especially China, which Birnbaum (2005, 2008, 2015) calls "the tailor of the world," to locales closer to Europe and the United States so that product can be transported to retailers in days rather than weeks (Bruce & Daly, 2004, 2006). Since data from China is not available, that particular claim cannot be tested, yet a general analysis can be performed to see if there is a significant investment difference between the regions. According to Bruce and Daly (2004, 2006), with the advent of "fast fashion," African and Asian nations, such as Nepal and India, became more attractive locales because of proximity and presence of land routes to the West European market. Additionally, evidence exists that since the early 2000s, Chinese textile producers started to locate out of China to strategic regions in Africa and Eastern Europe (Alden et al., 2008; Busse, 2010; Carmody & Owusu, 2007; Tull, 2006; Zafar, 2007) to be in a favorable turnaround competitive position to offer fast response to clients. That trend continues (Azmeh & Nadvi, 2013; Zhu & Pickles, 2014; Zhu, Pickles, & He, 2017). However, the model here cannot account for source of investment and no particular claims can be made with respect to China, which is a serious limitation. Additional limitations arise from the fact that not all nations with available textile pollution data have complete data for the other variables in the model. Table 4.2 indicates which nations are left out of the analysis and for which specific reasons.

The data patterns in Table 4.2 suggest that two of the nations – Turkey and South Korea – may be outliers. Both are developed, dynamic, and diversified economies. It is unclear how dependent on textile exports their economies are today. Therefore, an outlier diagnostic is performed using the "flag" command in STATA and the results suggest none of the nations in

Table 4.2 Countries Left Out of Analysis in Specific Years Due to Missing Data

Country	*Not Included in Analysis Due To:*
Albania	No CCD data; not included in any models[a]
Czech Republic	Not included for years 1991 and 1992 only, due to missing FDI data
Estonia	Not included in 1991 due to missing FDI data
Fiji	Not included in log-linear analysis because of missing percent of population living on less than $2 a day. Included in elasticity estimations, except for years 1996, 1997, 1999, and 2000 when FDI net inflows had negative values.
South Korea	Not included in log-linear models because of missing percent of population living on less than $2 a day. Included in elasticity estimations only for years 1998–01 and 2003–2005. For all other years FDI net inflows have negative values.
Lesotho	No CCD data; not included in any models[a]
Lithuania	Not included for years 1991 and 1992 only, due to missing FDI data
Madagascar	Not included for years 1991 and 2006–2008, due of missing FDI data
Nepal	Not included in log-linear analysis for years 1991–1995, 2000, and 2008 because of missing FDI data. Not included in elasticity estimations for years 2000, 2002, 2004, and 2006 when FDI net inflows had negative values.

[a] Additional tests excluding CCD were conducted in order to include this nation and the results were consistent with the general findings in the models to follow

the sample to be outliers. The results could be due to the fact that for some of the time series analyzed here, both Turkey and South Korea were classified as developing nations by the World Bank, based on GNI per capita. South Korea achieved developed nation status in 1997, while Turkey was still considered developing by the end of the time series in 2008.

During those decades both nations were also relatively dependent on textile production. According to UNCTAD's textile export dependence rank, based on textile exports as a share of an economy's total merchandise exports, in 2003 (the last year of the data analysis in the report published in 2005) Turkey was 4th in the world and South Korea 11th. In 2003 Turkey's percent of textile exports as a share of all merchandise exports was 11.7%, South Korea's was 5.6%, while Bangladesh's was 7.3%. These values could be related to economies of scale and technology capabilities. The same report also ranks each country's share of total world exports of textiles and Turkey is 3rd on the list with 4.2% in 2003, tripling its gross textile export values from 3.3 to 9.96 billion $US from 1990 to 2003. At the same time South Korea has notably divested domestically, with gross values of textile exports falling from 8 to 3 billion $US from 1990 to 2003. Yet, at that time South Korea was still a viable world exporter, accounting for 1.5% of all of the global textile exports in 2003, same as Bangladesh, sharing in the 10th ranking. For these reasons, both Turkey and South Korea are not treated as outliers.

With the available data, the general cross-sectional, time series econometric model is specified in Equation 4.1:

$$\begin{aligned}\mathbf{FDI}_{\mathbf{NETINFLOWSit}} = {} & \beta_1 \mathbf{H_2 0}_{\mathbf{TEXTILE}_{it-1}} + \beta_2 \mathbf{H_2 0}_{\mathbf{CHEMICAL}_{it-1}} + \beta_3 \mathbf{CDDamage}_{it-1} \\ & + \beta_4 \mathbf{AG}_{\mathbf{IMPORTSit}} + \beta_5\,\mathbf{WTO}\mathrm{it} + \beta_6\,\mathbf{GNI}\mathrm{it} + \beta_7 \mathbf{POP}_{\$2} + \beta_8 \mathbf{AFRICA} \\ & + \beta_9 \mathbf{EUROPE} + \beta_{10} \mathbf{ASIAandOceania} + \beta_{11} \mathbf{L}_{\mathbf{AMERICA}_{\mathbf{CARR}}} \\ & + \mathbf{FDI}_{\mathbf{NETINFLOWSit-1}} + \mathbf{e}_{it}\end{aligned} \tag{4.1}$$

where i = country, i = 1,2,…20 and t = time, t = 1,2,…18.

Lacking reliable data for environmental regulatory stringency, the proxies for environmental permissiveness – the water and air pollution metrics – are lagged by one time period in an attempt to control for endogeneity. It is unclear whether FDI causes pollution to increase – which is what the majority of the pollution haven literature has established – or if investors choose locales where pollution rises, seeing the rise as a signal of inadequate environmental regulation. Lagging the explanatory variables is an imperfect technique (Bellemare, Masaki, & Pepinsky, 2017) which in this model is even more problematic because the water pollution metrics do not show high degree of variability. However, the CCD values do and to formally test for endogeneity three endogenous covariate diagnostic tests are performed through single-equation instrumental-variable regression with: $\mathbf{FDI_{NET_{INFLOWSit}} = \beta_1\ H_20_{TEXTILEit-1} + \beta_2 H_20_{CHEMICALit-1} + \beta_3 CDDamage_{it-1}}$. $CCDamage_{it}-_1$ is specified as the hypothetically endogenous variable. Used in the diagnostics as instrumental variables for $CCDamage_{it}-_1$ are the following: (1) CCD lagged by two time periods, i.e., $CCD_{it}-_2$; (2) BOD emissions; and (3) the lagged values of overall carbon dioxide emissions – $CO_{2it}-_1$. A two-stage least squares (2SLS) test is performed with each of the instrumental variables, followed by a post-estimation test of explanatory variable endogeneity. In all three cases, both the Wu-Hausman F statistic and the Durban score p-values were above the critical value of 0.05.[2] Therefore, the null hypothesis that the variables are exogenous is accepted. $\mathbf{H_20_{TEXTILEit-1}}$, $\mathbf{H_20_{CHEMICALit-1}}$ and $\mathbf{CDDamage_{it-1}}$ are included in the analysis as exogenous independent variables, with the caution that with such limited data, the inability to sufficiently controlling for endogeneity is a major limitation.

A Hausman diagnostic indicates that a random effects test is the best fit for the analysis (Hausman, 1978). The Chi2(6) Hausman asymptotic assumption test coefficient is −1.72, which in absolute value does not meet the necessary threshold on below.05 to indicate a fixed effects test to be appropriate. Thus, Equation 1 represents a dynamic panel model with time-invariant explanatory variables, *i.e.*, the regional dummies that act as 5 fixed variables.

For random effects tests for panel data, the accepted statistical approach is to include a lagged dependent variable to control for autocorrelation as a function of a time effect, as well as endogeneity (Greene, 2008; Keele & Kelly, 2006). For FDI, this method is particularly appropriate because FDI net inflows depend on FDI stock (Chaudhuri & Mukhopadhyay, 2014; Mallampally & Sauvant, 1999), and in that way, the best predictor of future investment is already invested capital in a given nation. Thus, the model here is a hybrid. It is a dynamic model with natural "fixes."

To explain, fixed effects tests apply a "fix" on individual years and individual nations and thus, the calculations compare the variability among observations and then among times frames. Statistical difference is calculated between each observation and the first observation in the sample. Then the same is calculated for each year in the time series, comparing the difference to the first year of record. In the model here, that process occurs by default because of the categorical variables for region. Thus, the five fixed variables are the categorical classifications for Africa, Europe, Latin America, and Asia and Oceania, compared to the holdout regional code for South-East Asia.

Three tests are run in three time frames to examine the impact of policy change, which as noted, is the removal of the MFA. Model 1 evaluates the years 1991–1999 – the period under MFA directives. Model 2 evaluates the years 2000–2008 – the liberalization period of phase-out and removal of the MFA. Model 3 evaluates the whole time frame – 1991–2008 – to examine the magnitude impact of the removal of the MFA.

As noted, during the MFA years, FDI decisions were subject to the very strict rules of origin quotas. Investment flows changed when nations were able to change their quota restrictions through diplomatic ways. Birnbaum (2008) coins the phrase "chasing quotas" to explain the dynamic, detailing the trading of quota categories on the Hong Kong stock exchange.

The process to change these restrictions evolved during the GATT Uruguay Round from 1995 to 2000, when gradual protection for mostly Western textile producers was lowered and the goal to dismantle the MFA was set (Lu, 2012; Mlachila & Yang, 2004). The final agreement to dismantle the MFA was reached during the WTO Doha Round in 2001 when the Doha Development Agenda (DDA) was put in place. Although specific dates at the time were not set, negotiations on the subject commenced. Thus, even though the MFA did not formally end until January 2005, future facility investment decisions would have been made under unrestricting rules of origin platform. Therefore, it is of interest to examine if the *formal* agreement to end the MFA impacted FDI, rather than the calendar end of the system. For these reasons, the goal is to evaluate if given a choice, producers faced an incentive to choose possible pollution havens.

Table 4.3 presents the results of Model 1, Model 2, and Model 3.

The results show no significant findings in the 1990s, except for a positive relationship denoted at the 90% confidence level between $CCD_{it}-_1$ – the proxy for environmental permissiveness – and FDI net inflows. This relationship increases in magnitude during the 2000s to a level that impacts the whole time series – the years 1991 to 2008. The interpretation is that investors increasingly favored locales with evidence of environmental permissiveness. In the 1990s, none of the other independent variables are significant. In the 2000s, things change. Rising effluent pollution levels from textile and chemical production become significant predictors of increasing FDI, as do the national wealth proxies, and the regional dummy variables.

Table 4.3 Influence of Pollution on FDI Net Inflows, [a]Nations Most Dependent on Textile Export

	Model 1	*Model 2*	*Model 3*
	1991–1999	*2000–2008*	*1991–2008*
$H_20_{TEXTILE_{it-1}}$ (Water Pollution, Textile Industry)	0.94 (4.22)	63.04** (23.13)	17.95** (5.97)
$H_20_{CHEMICAL_{it-1}}$ (Water Pollution, Chemical Industry)	13.52 (12.13)	77.65** (25.60)	48.08*** (12.73)
$AG_{IMPORTS\ it}$ (Agricultural Raw Material Imports)	−89.47 (54.85)	−769.87** (262.41)	−324.37** (109.78)
CDD_{it-1} (Carbon Dioxide Damage)	4.16e−08^ (4.61e−07)	2.82e−06*** (8.24e−07)	1.55e−06*** (4.53e−07)
WTO (WTO Membership)	−44.42 (172.77)	−4.43 (118.70)	−71.64 (130.97)
GNI_{it} (GNI per Capita)	0.08 (0.08)	0.06 (0.08)	0.05 (0.04)
$POP_{\$2}$ (People Living on under $2 per Day)	−4.03 (5.04)	−55.67** (18.42)	−17.70*** (5.27)
$AFRICA_i$ [b]	243.98 (286.43)	1307.95* (557.40)	590.82** (214.21)
$EUROPE_i$ [b]	−227.24 (306.68)	598.02 (650.72)	81.67 (276.70)
ASIA and $OCEANIA_i$ [b]	628.27 (509.48)	5594.47** (1778.79)	2166.57*** (649.35)
$L\_AMERICA_{CARR}$ (Latin America & The Caribbean)$_i$ [b]	−36.07 (170.47)	−1904.37* (836.19)	−346.73 (248.54)
$FDI_{NET\ INFLOWSit-1}$	0.87** (0.27)	0.55*** (0.13)	0.73*** (0.07)
Constant	49.41 (471.79)	−692.63 (616.11)	−217.36 (336.98)
Model Details			
Probability > F	<0.0001	<0.0001	<0.0001
R-squared [c]	0.66	0.76	0.74
Observations	118	109	227

[a] **Dependent Variable:** FDI net inflows – the overall balance of foreign assets to liabilities in a country measured in millions of current US dollars.

[b] Holdout region = South-East Asia

[c] Wald Chi two-tailed tests, where NS = not significant, ^ p<0.10, * p<0.05, ** p<0.01, *** p<0.001

Clustered standard errors in parentheses

From 2000 to 2008, FDI inflows significantly *decreased* in the Latin American textile exporters by an average of $US 1.9 billion and *increased* in Africa by 1.3 billion $US and in Asia and Oceania by over 5.5 billion $US, as compared to flows into South-East Asia. The magnitude of the increases in FDI inflows into Arica and Asia and Oceania is such that it defines the overall trend of the entire time series for FDI into those regions, as shown in Model 3. This finding provides empirical backing for Bruce and Daly (2004, 2006) who posit that because of fast fashion's shortening lead times, locating production facilities in African and Asian nations with land routes to Western Europe, would become important for apparel manufacturers. It is also congruent with works that analyze the continuing growth of textile production clusters in newly important strategic locations because of their proximity to Western markets (Au & Wong, 2007; Kilduff & Chi, 2006; Morris, Plank, & Staritz, 2016).

The results also show that FDI inflows significantly increased in nations where poverty rates were decreasing and where agricultural imports, as compared to imports from other sectors, were also decreasing. In comparative works, such a trend denotes industrialization growth, as countries import more technologies, capital equipment, and durable goods (Vamvakidis & Dodzin, 1999). The interpretation is that FDI flows significantly increased in the relatively wealthier nations in this sample. These marginal interpretations should be viewed with caution as the estimates they are based on are imperfect proxies, as already explained, based on World Bank assessments. Yet, with the limitations of the data estimates, the low variability of the numbers, the use of a lagged dependent variable, and lagging the explanatory variables to minimize time effect and autocorrelation, the presence of significant relationships must be noted.

Those relationships remain significant in robustness checks, employing different proxies for overall environmental permissiveness on FDI, as well as additional elasticity estimations where percent of population living on less than $2 is left out of the regressions in order to include the nations missing this data, which are Fiji and South Korea. Included instead as a poverty proxy, is the natural log of GNI per capita. In the robustness checks the CO_2 damage variable was replaced by estimates of mineral depletion, overall toxic particulate emission damage, discreet CO_2 emissions in metric tons, as well as oil equivalency, and combustible renewables and waste, all part of the *World Bank Development Indicators* database. The significant impact of these proxies on FDI net inflows varied (possibly because of missing values); however, the impact of the water and textile pollution variables, as well as the wealth proxies, remained consistent. An additional robustness check is presented in Table 4.4, where FDI net inflows are replaced with the size (and growth over time) of the "clothing and textile" sector, as the best available proxy for sectoral trade flows. Values also come from the World Bank's *World Development Indicators* data catalog.

The World Bank calculates the size of the local apparel sector as percent of all manufacturing. Therefore, congruently with the previous tests and for ease of interpretation, the results are presented as elasticity calculations. The robustness checks show the same patterns of significance as the FDI models and for brevity. Table 4.4 includes only the entire time frame of the study. As with FDI inflows, the results show sectoral growth to be directly impacted by rising water pollution. Additionally, and consistent with

Table 4.4 Elasticities of Pollution on Size of Local Clothing Industry, Nations Most Dependent on Textile Export, 1991–2008

	Model 2[a]	*Model 3*
	2000–2008	*1991–2008*
$H_2O_{TEXTILE_{it-1}}$ (Water Pollution, Textile Industry)	0.02***	0.009*
	(0.006)	(0.004)
$H_2O_{CHEMICAL_{it-1}}$ (Water Pollution, Chemical Industry)	0.03***	0.03**
	(0.009)	(0.008)
$AGIMPORTS_{it}$ (Agricultural Raw Material Imports)	–0.05	–0.003
	(0.04)	(0.04)
$logCCD_{MIL_{it-1}}$ (Carbon Dioxide Damage)	0.36***	0.17***
	(0.08)	(0.05)
$logGNI_{it}$ (GNI per Capita)	0.77***	0.44***
	(0.19)	(0.11)
$FDI_{NET\ INFLOWS_{it-1}}$	0.49***	0.67***
	(0.14)	(0.06)
Constant	–5.87***	–2.95**
	(1.78)	(0.99
Model Details		
Probability > F	<0.0001	<0.0001
R-squared [b]	0.81	0.84
Observations	107	222

[a] **Dependent Variable:** FDI net inflows – the overall balance of foreign assets to liabilities in a country measured in millions of current US dollars.
[b] Wald χ^2 two-tailed tests, where NS = not significant, ^ p<0.10, * p<0.05, ** p<0.01, *** p<0.001
Clustered standard errors in parentheses

Table 4.5 Elasticities of Pollution on Size of Local Clothing Industry, Nations Most Dependent on Textile Export, 1991–2008

	1991–2008[a]
$H_2O_{TEXTILE_{it-1}}$ (Water Pollution, Textile Industry)	0.23*** (0.09)
$H_2O_{CHEMICAL_{it-1}}$(Water Pollution, Chemical Industry)	0.44*** (0.13)
$AG_{IMPORTS\ it}$ (Agricultural Raw Material Imports)	0.45* (0.17)
$logCCD_{MIL_{it-1}}$ (Carbon Dioxide Damage)	–1.06** (0.33)
$logGNI_{it}$ (GNI per Capita)	1.90** (0.61)
$TEXTILES_{AS\ \%\ OF\ TOTAL\ MANUFACTURING_{it-1}}$	0.75*** (0.08)
Constant	–18.37** (6.27)
Model Details	
Probability > F	<0.0001
R-squared [b]	0.95
Observations	253

[a] **Dependent Variable:** Percent of Clothing and Textile sector as percent of total manufacturing.
[b] Wald Chi two-tailed tests, where NS = not significant, ^ p<0.10, * p<0.05, ** p<0.01, *** p<0.001
Clustered standard errors in parentheses

previously discussed development research, Table 4.4 and Table 4.5 shows that the apparel sector is a larger percent of overall manufacturing in nations with lower air pollution levels and higher agricultural import ratios, but with notable GNI per capita growth. These results capture the fact that apparel production is concentrated in fairly underdeveloped nations – as denoted by lower CO_2 emission damage and higher agricultural imports – that are experiencing economic growth significant enough to result in increasing GNI.

Using the values of the pollution approximations from the World Development Indicators in absolute terms at the national level is problematic. Employing them as estimates of outcomes of overall economic activity in a country and its environmental regulatory stringency, without the ability to disaggregate either, can lead to ambiguous results. In the model here, the inability to control for textile FDI specifically and any national variations in environmental stringency laws are serious limitations. Therefore, the above results must also be interpreted with the caveat that they are dated, exclude the main textile and apparel production nations China, India, and Pakistan, and cannot track source of FDI. Future studies should make efforts to investigate sources of textile FDI, both in terms of investing nation and corporate entity, to better assess if MNCs from relatively wealthier nations, with stronger environmental stringency laws, are still actively seeking pollution havens. If so, research should be focused on understanding what specific regulatory loopholes in pollution mitigation are offered by recipient governments.

In summation, the finding here show a change in FDI net inflows that started with the phase out the MFA toward nations with rising pollution levels, proximity to main retail markets, evidence of increasing industrialization, and decreasing poverty rates. These results are interesting in the context of the argument that, in general, apparel producers chase cheap labor (Anner, 2009, 2011) and the cheapest workers are found in the poorest countries (Braconier, Norbäck, & Urban, 2005). That very well could be the case, however, as Birnbaum (2008) shows, overall labor costs (at that time) were approximately 4 % of integrated apparel expenditures. With a focus on the direct costs of textile production, which Birnbaum (2008) shows to be three times the amount of overall labor costs, the results here suggest that for textile producers, cheaper labor may not have been essential.

The main implication of the significance of the wealth proxies is that cost-cutting admonitions, in terms of wage rates, are not sufficient to explain the site selection choices of textile producers. It appears that site selection is more directly impacted by capacity, environmental permissiveness, and proximity to retail markets. The documentary film *The World According to H&M* thoroughly covers the incentives to locate apparel production facilities in impoverished developing countries, which include weak governance, lack of labor and environmental regulation, and willingness of local governments to tolerate labor and environmental exploitation (Maurice & Hermann, 2017). The film explains that the lack of government oversight on operations is a function of pressures from apparel conglomerates. The same dynamic is tracked in *Zara: The Story of the World's Richest Man*, in the analysis of a cluster of textile factories in Tunisia that produce exclusively for *Inditex* (FilmsMedia Group, 2016). What the results here add to the story is that this location trend started almost two decades ago, with the removal of the MFA.

It appears that the creation of pollution haven clusters in the textile sector to be a function of a change in the main industrial model of fashion economics and the consequent policy change implemented to accommodate the resulting sectoral growth. Namely, the evidence suggests that with the proliferation of fast fashion, apparel producers spearheaded the dismantling of the geographical restrictions governing the international trade of apparel. Codified under the rules of origin provisions of the MFA, the geographic restrictions were phased out gradually from 1999 to 2005, with some protections in place until 2008. The results here indicate that the phase-out period changed FDI inflows in the nations most dependent on textile exports, *de facto* creating pollution haven clusters in strategic supply chain locales.

Whether the removal or the MFA was orchestrated to allow for the creation of pollution havens in textile manufacturing is a question that merits further investigation. The evidence presented here would allow for such a logical conclusion, yet it would be over-reaching. Positing that fashion conglomerates willfully created a system of environmental injustice is a damning accusation. However, for a $3 trillion industry, governed by a few thousand decision makers (Anguelov, 2015) with some of the most powerful global lobbyists (Baffes, 2011; Barnett, Grolleau, & Harbi, 2010), it would be naïve to accept that the perverse incentives behind the creation of textile pollution havens happened unintentionally. As the literature examined here on the long and complicated negotiation process to remove the MFA shows, Western apparel producers – those that in the first place created the MFA in the 1960s for their own market protection – engineered the removal of the MFA in the mid-1990s. Their market incentives changed with the advent of fast fashion from a business model of localized commerce to one of globalizing operations in manufacturing and retail. The MFA's restrictiveness impeded efficient global sourcing and strategic international product positioning.

Those are the reasons given by professional and academic analysts for the removal of the MFA. No one has raised the question whether behind the change of policy was also a quest to evade environmental regulations. It is hard to imagine that those lobbying for the removal of the MFA would not have identified the fact that its market liberalization provisions would make it easier to locate in nations with permissive environmental governance.

The issue needs to be investigated further. It has implications for environmental regulation analysis because it raises questions on policies that may create systemic perverse incentive problems. For industrial sectors where pollution mitigation technologies are lacking, would manufacturers respond to increasing capacity demands by lobbying for a change in trade policy toward liberalization? In that way, would the well-established and accepted policy of trade liberalization lead to perverse incentives? These are questions much discussed in works across academic fields. They are not easy to answer and data collection and operationalization quests must continue.

The analysis here shows that in apparel production, as sale volumes keep increasing due to rising demand in a globalizing market, environmental regulation, which is subject to national discretion, can be more negatively impacted in developing countries with higher industrialization growth rates. It is understandable, as well as much discussed in previous studies on industrial upgrading, that in the developing world in particular, governments will prioritize economic growth. The problem is that this growth back then, and to an extent today, is most often a function of corporations from the developed world.

Although the number of global apparel conglomerates from the developing world is increasing, as explained in Chapter 2, the main fashion players are not only from the developed world, but, as is the case of *H&M*, from nations with cultures that prioritize environmental stewardship, or at least posit so. These corporations are well versed in environmental regulation stringency and compliance. The results here indicate that, when given the option, such corporations can choose to invest in nations that allow pollution to increase. The option, in this case, is a policy change of trade liberalization – the ending of the MFA and with it, the removal of a restrictive and protectionist system. The reasons why are vested in the legacies of modern policies of economic growth. They prioritize growth above all, with specific policy tools to stimulate consumption.

Notes

1 Definition available at https://datahelpdesk.worldbank.org/knowledgebase/articles/114954-what-is-the-difference-between-foreign-direct-inve

2 Specifically, with CCD_{it-2} as an instrument for CCD_{it-1}, the Durbin score p=0.98 and the Wu-Hausman F p=0.98. For the test in which BOD is an instrument for CCD_{it-1}, the Durbin score p=0.57 and the Wu-Hausman F p=0.58. Finally, with $CO2_{it-1}$ as an instrument for CCD_{it-1}, the Durban score p=0.23 and the Wu-Hausman F p=0.24.

Chapter 5

The Policies of Consumerism

Hyper Economic Growth and Environmental Action

Since the 1990s development and economic analyses have been focused alleviating poverty in the developing world through stimulating economic growth (Rodrik, 2003; Szirmai, 2012). The end of the Cold War, the technological advances in mass communications, and the globalization of commerce and governance ushered a coordinated and targeted, era of policy. Although much of the academic world has focused on the growth component of the private sector, that growth itself is a function of the liberalization policies that enabled private firms to proliferate internationally. Often however, the analysis stops with this explanation – it was policies of liberalization, meaning freeing market access and easing business restrictions – that led to what has been termed "hyper globalization" (Haynes, 2003; Perraton et al., 1997; Perraton, 2019). Timmer et al. (2016) focus on the role of firms' reliance on global value chains (GVCs), defined as the relocation of production to low-wage nations. In the fashion industry, the work of Gerry Gereffi (1999) is best known and most cited for its thorough analysis of apparel GVCs.

Timmer et al. (2016) simplify the outcome of MNCs reliance on GVCs with the term "containerizaion" of production and posit that it is a result of the end of the Cold War, the admission of China into the WTO, and the advances in transportation technology. As a matter of fact, it is the "hyper economic growth" of China and in particular its history during the hyper globalization years, which the authors claim were 1990 to 2008, lasting until the onset of the 2008 recession, that is most analyzed by development scholars. Jin (2020) tracks the Chinese economic growth rates and concludes that their high and sustained trajectory started in the late 1970s. Market liberalization, with the admission of China into the WTO, other policy innovations such as the removal of the MFA, and technological advancement spearheaded the hypergrowth rates, not only in China but throughout the Global South.

This model of rapid industrialization was first tested in Japan in the 1950s and 1960s (Chen, 1997), and on the shoulders of, so to put it, the textile industry (Anguelov, 2015) repeated in the "Asian Tigers," the group of nations that today comprise Association of Southeast Asian Nations (ASEAN), and were so described in the 1980s and 1990s. These nations moved rapidly from the agrarian to manufacturing stage of national economics, relying heavily on, as is the term "export-led" growth (Shan & Sun, 1998; Shirazi & Manap, 2005.

This export-led growth depended on the cheap labor "value" of GVCs, giving employment to local populations that could easily leave subsistence farming for low-skill, mass-production employment. The problem is that the pursuit of export-led growth, albeit understood in the context of lowering and ending extreme poverty, has significant, if not catastrophic, environmental outcomes. In China, in particular, the issues are well discussed and documented and stem from lack of governance of pollution regulation and mitigation (Wen & Li, 2007).

This proliferation needed fuel. A developing nation marred in poverty does not become an attractive place to build new facilities overnight. As Chapter 4 explains, in fashion economics, the "winning," if one can call them that, nations for apparel production investment, were, and still are, nations with not just cheap labor, but reliable infrastructure, where local governments have a strong economic say (FilmsMedia Group, 2016; Maurice & Hermann, 2017). Strong centrally planned industrial policies of building such infrastructure are the priority under which political and economic powers converge.

It is this fact that leads to an environmental paradox. It is apparent in apparel production that economic profit in the short-term trumps ecological considerations. To understand the paradox and its pervasiveness, one must examine why it persists from a policy prospective. The interplay between economic growth policies and environmental policies has been examined by the academy from across disciplines since the 1960s. Yet, few viable solutions are offered. In the case of fashion economics, none can be deployed.

Since the onset of modern environmental policy in the 1960s, the tradeoffs between economic development and environmental protection continue to be subject to rigorous empirical research. As the social activism around climate change increases, it appears that all academic disciplines are contributing to the analysis. Yet, there is not much agreement on prioritizing conservation vs. production.

The hard truth is that production *is* pollution. Therefore, lowering pollution would mean lowering production and that is not something that is economically, socially, or politically viable. Hence, the discussion and academic research focus on trade-offs, regulation, and the development of new, now referred to as "green" or "carbon neutral," technologies.

The foundation of such bodies of research is based on the formalization by Ehrlich and Holdren (1971) of "I" – environmental impact of development – as the interaction of population, affluence, and technology. It has become known as the IPAT equation: I = P*A*T, where P is population, A is affluence, and T is technology. The majority of analyses use the equation at the national level, under the assumption that as nations become wealthier, their levels of affluence and technological capabilities increase, therefore, they are able to lower their environmental impact, respective to population size.

Much discussion has been generated around the assumptions of the equation because two of its factors – "Affluence" and "Technology" are ambiguous (DeHart & Soulé, 2000; Johnson & Villumsen, 2018; MacKellar et al., 1995; Soule & DeHart, 1998). Furthermore, at large, they are applied on "per capita" basis, which creates an equity problem. The "per capita" notation assumes an "equally-borne-by-all" scenario of environmental damage, which is not the case because of "super polluters" (Ehrlich & Ehrlich, 2010; Lane, 2017). Super polluters are industries, and their respective facilities, for which little to no technology exists for purification or filtration of effluents and emissions. This persistent problem is behind the growth of the literature on environmental justice (Collins, Munoz, & JaJa, 2016; Lynch, McGuire, & Smith, 2020; Schlosberg, 2009).

With case study analysis from the developed and developing world, the environmental justice works show that on a national level, relatively more polluting industrial activity is disproportionately impacting the poor and vulnerable. In the developed world, these are often communities of people of color. In the developing world, the problem is exacerbated in nations marred in extreme poverty. In both the developed and developing world, facilities are locating where little oversight from community can be expected and where governance is poor. Such communities do not efficiently increase their levels of affluence and therefore, the assumption that the interplay of affluence and technology would decrease environmental impact falls short. In fact, consumption-based studies have shown that, as nations increase their level of "Affluence," moving from economies of "subsistence agriculture" to "rural" to "urban" industrialization, ecological problems increase (Lambin & Meyfroidt, 2011; Weinzettel et al., 2013; Verones et al., 2017).

During those stages of the industrialization process, countries become more reliant on exports and also engage in trade liberalization political action in order to increase international market access. In this way, international trade also contributes to environmental exploitation, which the literature has covered well. For decades scholars have written about the dynamic, and not much has changed. The few works that find evidence of a positive environmental impact from the interaction of national wealth and international trade show the outcomes to be isolated, and mainly in the developed world (Inglehart, 1995; Apergis & Ozturk, 2015; Antweiler et al., 2001).

The problem is that modern economic thought relies on the basic idea that trade increases wealth and wealth is assumed to be the necessary condition for environmental action. The logic is that, as they get wealthy, nations would reach a certain level of affluence when citizens would be willing to invest in healthier environments rather than personal economic gain. The question is when? How wealthy must citizens become to stop prioritizing self-enrichment? Answering such questions honestly would challenge the positive assumption that wealth growth would create, as Inglehart (1995) puts it, "social conditions" for environmental action. The hope is that such action would be in the form of legislation. This is the main outcome of the process the Environmental Kuznets Curves (EKC) hypothesis describes. The problem is that it is based on the birth and proliferation of environmental protection laws in the industrialized West. Even there, it is obvious environmental problems persists, and therefore, the EKC has also received much criticism, questioning its theoretical merit (Dinda, 2004; Gill et al., 2017; Perman et al., 2003; Tisdell, 2001).

Environmental Kuznets Curves Under Globalization

The majority of works that find evidence of EKC dynamics at the national level analyze air pollution. Most often studied air pollutants are Sulfur Dioxide (SO_2) and Carbon Dioxide (CO_2), a fact due to their readily available values, which have been collected and recorded by governments since the 1960s. Researchers have observed decreases of their emissions as a function of renewable energy generation, supporting the assumption of the importance of both the wealth and technology factors in the IPAT equation (Inglehart, 1995; Apergis & Ozturk, 2015; Jebli, Youssef, & Ozturk, 2016). However, such works have received criticism for focusing on national level air pollution (Stern, Common, & Barbier, 1996; Gill et al., 2017), while studies on global CO_2 footprints have shown no EKC evidence (Pablo-Romero & Sánchez-Braza, 2017). That is the case because the hyper economic growth wave is still on-going, mainly outside the industrialized West.

As previously explained, that growth, which is also subject to growth in the volume of international trade, is not resulting in viable environmental policies. Even in the developed world, there is evidence that international trade, although an aggregate generator of national wealth, does not *per se* lead to an efficient redistribution of that wealth toward creating EKC incentives. For example, tracking Portugal's industrial development from 1971 to 2008, Shahbaz et al. (2015a) find trade openness to have had no impact on pollution.

At present, the EKC is stretched to serve scholars on two theoretical fronts – that of public health hazards (Al-Mulali et al., 2015; Rasli et al., 2018) and that of climate change (Apergis & Ozturk, 2015). The problem is that they are different in kind and in type. Public health scholars focus on evidence of environmental protection from substances known to be directly harmful to human health. The quest is to establish a connection between pollutants and a change in policy to protect the public from the harmful effects of these pollutants. Climate change queries, which dominate the EKC literature, look at aggregate pollution levels under the more relaxed assumption that eventually, *aka*, in the "long run," those pollution levels would lead to harmful outcomes for both humans and the planet in general. Few, if any works, dare to estimate when such an eventuality can reach a tipping point, which would cause a policy change.

Admonitions have also been made that richer nations can displace their own pollution emissions through importing, rather than producing, toxic goods from poorer nations (Arrow et al., 1996; Stern et al., 1996; Rothman, 1998; Dinda, 2004). Chapter 4 in this book suggests this could very well be the case in textile manufacturing. Rasli et al. (2018) specifically show that trade openness increases environmental degradation. Similar to the findings in Shahbaz et al. (2015b) for Portugal, Rasli et al. (2018) observe the link in 36 nations at different levels of development from 1995 to 2005. This an important time frame of rapid economic growth, increasing levels of international trade, as well as relatively higher rates of technological and telecommunications innovation (Davis, 2000).

All these factors created an environment in the 1990s in which "spatial costs" to trade and technology transfer were significantly lowered (Neary, 2009). Spatial costs is the collective term describing the direct and indirect (transaction) costs of transporting and communicating across large distances. Pulitzer Prize winner Thomas L. Friedman (2006) described the gradual decrease of spatial costs so eloquently in his book *The World Is Flat: A Brief History of The Twenty-First Century*, that the words "the world is flat" became an allegorical phrase used by economists and trade analysts.

Low spatial costs, the end of the Cold War, and general technological advances, accelerated the rate of international trade to such levels that the concept of "internationalization of the production function" emerged to describe productivity based on global supply chains (Kleinert, 2001; Merino, 2004). International trade became global trade and the cornerstone of modern economics and politics. This internationalization led to, as is the term, global economic "integration," incentivizing intergovernmental cooperation (Marsh & Sharman, 2009; Simmons & Elkins, 2004).

It is this dynamic that fueled the notion that wealth created via trade can lead to social benefits, linking the assumptions of wealth creation and environmental stewardship of the EKC to international trade (Perman et al., 2003). Specifically, Antweiler et al. (2001) argue that it is "additional income" from trade that leads to investment in new, low-polluting technologies. Examining SO_2 levels, the authors find "scale effects" among trading partners, where as a function of trade, production increased. Scale effects is the economic term of increasing production while decreasing the cost of production as a function of innovations. Antweiler et al. (2001) observe that SO_2 net emissions decreased among trading partners, and therefore, conclude that it was profit from trade that created the necessary capital for investment in more efficient technologies. However, using a similar sample of countries, De Bruyns (1997) finds SO_2 emission reduction to have been achieved not as a function of trade, but of policy. Specifically, it was the agreement of trading partners to sign to The Convention on Long Range Transboundary Air Pollution (LRTAP). Grether, Mathys, and de Melo (2009) also find evidence of economies of scale growth as a function of trade congruent with a decrease in SO_2 emissions. However, the decrease is insignificant. The authors show that from 1990 to 2000 among the 62 nations responsible for 76% of global SO_2 emissions, emissions only decreased by 10%.

As already explained by the works on hyper economic growth, such numbers reflect a time frame of exponential increase in trade, while they also capture another fact – there is an available and fairly well-regulated environmental mitigation option for the lowering of SO_2 emissions. It is referred to as "scrubber" technology and for decades governments, at both the national and local levels, have implemented policies to mandate their use (Srivastava, Jozewicz, & Singer, 2001; Popp, 2006; Taylor, Rubin, & Hounshell, 2005). The technologies are also fairly efficient, being able to capture over 90% of SO_2 (Grether, Mathys & de Melo, 2009; Hrdlička & Dlouhý, 2019).

Such works reflect a very visible social driver of environmental technological and policy deployment – the knowledge of the dangers of air pollution. One can see it; one can easily understand its pulmonary impacts; one can readily support its regulation politically. With other forms of pollution where the particulate articles are harder to discern (one cannot see Mercury in water), negative health effects are not as intuitive. The social discourse on their direct impact on climate change is less vociferous and attention toward a need of environmental action may not be as strong. In the case of water pollution from the fashion industry, social outcry is just a few years old. Therefore, technologies for filtration are still developing, with few commercial options available at present.

The major limitation of EKC research is lack of focus on social drivers that build awareness and knowledge to prioritize environmental action. Wealth is not sufficient. Knowledge is needed. One can argue that wealth can increase investments in creating knowledge; therefore, education at large is the key factor that would lead the public to show political support for EKC legislation. However, researchers have shown that while education can change environmental awareness, it may not necessarily lead to a change in pro-environmental behavior of individual citizens (Dietz et al., 1998; Franzen & Meyer, 2009; Kollmuss & Agyeman, 2002; Stern, 2000). This outcome is the definition of the tragedy of the commons (Dietz et al., 2003).

The tragedy of the commons describes how ecological degradation continues because, while aware of the problem, individual citizens fail to adjust their behavior. Rather, they expect others to act on their behalf or for government to provide solutions. This is why, as explained in Chapter 3, there is little evidence that at large, fashion consumers are changing the way they shop, wear, and discard their clothes.

It is not to say that environmental regulation is moot and void. It has been effective in reducing pollution (Ringquist, 1993). Therefore, EKC analysts should use a modified version of the IPAT equation to include the importance of policy. So far, the body of research has developed and continues to develop without a unified, global governance impetus behind it.

Today the academy is urged to tackle the reconciliation of its legacy of economic development research with the paradox that fuels the escalation of pollution. That paradox is the promotion of consumption, and specifically increasing the levels of aggregate consumption. In effect, it is all we know. From the development of the Keynesian school of economic thought, based on the works of John Maynard Keynes (1936), market-based development policies have relied on tools to increase consumption. Not until fairly recently has there been a discussion on seriously addressing the social costs of consumption.

The Consumption Paradox

Keynesian economics follows this logic: Via, a combination of, as are the terms, "monetary policy" and "fiscal policy" tools, consumption is to be stimulated by increasing demand in order to boost economic growth. Demand is increased by freeing up moneys for consumers that would otherwise be diverted from their income through taxation toward the provision of public goods. Keynes argued that in times of economic hardship, government should use its funds, accrued through taxation to, in effect, give back to its citizens so that they could increase their consumption. As such, monetary policy is targeted at businesses by lowering interest rates, so that businesses can borrow more money and expand production, therefore increasing their demand for employment. Fiscal policy is lowering taxes, as well as increasing government spending via varieties of stimulus options aimed at the consumer, in order to make citizens spend more money.

Keynes was writing on the heels of the Great Depression, redefining the role of government intervention in the private market to offer solutions for tackling depressions. Back then, nobody worried about pollution or what a government intervention in the market to stimulate consumption would do to the environment as a function of such consumption. The problem is that for decades to come, Keynesian policies would become the backbone of economic development for the major industrialization trajectory of humanity, unadjusted for environmentalism. From employing them to recover from World War II in the establishment of the Bretton Woods institutions – the International Monetary Fund and the World Bank – it was Keynesian logic that was put into policy. Simply, that it was the role of government to stimulate consumption to generate economic growth. That assumption defined, and to this day, drives the modern, hyper-globalized, hyper-growing economy of the world.

John Maynard Keynes was one of the three major masterminds that created the Bretton Woods supranational governance organizations, as the main vehicle of policy diffusion in economic learning and international finance. Particularly for the developing world after World War II, as it was first decolonized, then started to move toward industrialization, it was conditionality for international trade access and integration with the wealthier Western markets that these institutions put forth. The aim was to reach a noble goal – the reduction of extreme poverty by promoting industrialization (Boockmann & Dreher, 2003; Stiglitz, 2003). Working with the United Nations, established a year after the Bretton Woods institutions in 1945, these three governance structures shaped the global economic path of market-based, but government-controlled capitalism, as an antidote to the government-owned and managed socialist economic model of the Eastern Bloc.

From the industrialization decades of the 1950s, 60, 70s, and 80s, to the knowledge-based economy of the 1990s and 2000s, to todays "creativity economy," global trade and diplomacy policy has been based on increasing consumption. Chapter 1 of this book tracks how this consumption grew in the fashion industry. It also grew in every sector.

From food, to shelter, to leisure, today's humans are economic machines, constantly stimulated to consume more. Their livelihood, the prosperity of their nations, and the quality of life in their households are defined by how much they spend on consumption. Therefore, it is unprecedented that in 2018 the United Nations itself finally admitted that this trajectory of ever-increasing consumption is not sustainable. As part of the Sustainable Development Goals platform, on the UN's webpage is the proclamation: "If we don't act to change our consumption and production patterns, we will cause irreversible damage to our environment."[1]

But can we? Can we consume less? Would we consume less? For two decades now, consumer behavior as a function of concerns about climate change has been studied across disciplines. The problem is that even such research is mainly focused on spending. Marketing analysts have been primarily interested on whether consumers would engage in "substitution," as is the term, to buy "greener" products. From willingness to purchase electric vehicles (Hidrue et al., 2011) to demand for LEED[2] certification in home energy improvements (Rastogi et al., 2017), to putting solar panels on homes (Dastrup et al., 2012), scholars brandish the need to support the growth of green industries through consumption. The implication is that the reader should buy the products of those emerging green industries.

Little is said that the technologies behind the mass-market production, as well as the recycling and reclaiming efforts of such products, are still very toxic (Binnemans et al., 2013; Shin, Kim, & Rim, 2019). The production is based on extraction of specific minerals, ores, and metals, commonly referred to as "rare earths," (Hedrick, 1995), as well as the adequate petroleum to produce the energy for carbon-intensive extraction, refining, and transportation. The extractive processes employ toxic chemical methods, where poisonous byproduct leaches into soils and water, much like in textile dyeing and finishing (Ganguli & Cook, 2018).

Although such research exists, aimed at scholars of cleaner production, the aggregate sustainability attention of the academy is directed toward the emblem of climate change – CO_2 emissions. All other pollution is neglected at large in the general societal discourse on climate change. Even the term "climate change" itself denotes a focus on atmospheric, i.e., air toxicity. Therefore, it is of little surprise that consumers do not consider many of the products they buy to be problematic because they do not envision them as a source of CO_2 emissions. One sees cars, planes or coal-fired power plants as the culprits. Climate activist Greta Thunberg famously refuses to fly. However, she never says anything about not buying a new smart phone for example, arguably among the most polluting products today. Much research is emerging on the toxicity of extraction of the necessary rare earths for smart phone manufacturing, as well as the poisonous and dangerous amounts of "e-waste" from the frequent replacing of smart phones, and other devices, that are piling in unregulated landfills in poor nations, which are becoming the dumping grounds of the world (Garlapati, 2016; Nnorom & Osibanjo, 2008; Robinson, 2009). If anything, Thunburg proudly showcased the use of her smart devices as she tweeted and live-streamed her journey by yacht to the United Nations Climate Summit in 2019.

In short, even as we, the consumers, are bombarded with climate change activism messages, we make choices on how to interpret what we hear through our consumption. Thunberg tells the world she chooses not to fly, but not if she chooses to not consumed fresh produce flown daily into Sweden, like bananas for example, available in every Swedish market year-round. With respect of fashion, since the early 2000s, ecological stewardship through consumption messaging has shown that even though fashion buyers are worried about the environment, they are at large, unwilling to make changes in their consumption habits (Bray, John, & Kilburn, 2011; Manchiraju & Sadachar, 2014; Park & Lin, 2018). Even when there is evidence of such behavior, it is in the context of minority purchases. Bly,

Gwozdz and Reisch (2015) aptly use the phrase "sustainable fashion consumption pioneers" to describe the market segment of their study's population of interest – those fashion shoppers who actively change their buying behavior. The authors track the motivational factors of consumers who purchase fewer pieces of higher quality items, stop shopping altogether, buy consignment only, or sew and mend old garments. These buyers redefine style and are less focused on traditional understanding of fashion, opting for forming a social subculture. At large, such social formations are still quite small (Cho, Gupta, & Kim, 2015; Harris, Roby, & Dibb, 2016).

Notes

1 https://www.un.org/sustainabledevelopment/wp-content/uploads/2016/08/12.pdf
2 LEED stands for Leadership in Energy and Environmental Design.

Chapter 6

Sustainable Fashion Legislation

An Analysis of Emerging Networks in Global Governance

The Need for Formal Governance as a Function of Industrial Innovation and Growth

To incentivize innovation toward developing more viable "circular" production, governments are taking action. Legislation is evolving at the local, national, and supranational level to usher in a system of transparency in production and supply chain management. Such legislation is a function of the legacy of a growing body of research on transparency in fashion supply chain initiatives (Bhaduri & Ha-Brookshire, 2011; Egels-Zandén, Hulthén, & Wulff, 2015; Khurana & Ricchetti, 2016). The findings sum up the fact that such initiatives are, at large, voluntary and taken at the firm level. The benefits accrue to participating firms to help them build "eco-cache" with customer bases, which improves competitiveness for brands in a culture of growing social support for sustainability, however one defines it. The research notes that the incentives to participate bear higher operational costs. Therefore, few producers choose to make viable policy commitments toward a systemic increase of such costs in an industry that competes on price (Birkey et al., 2018; Köksal et al., 2017; Perry & Wood, 2019).

To tackle the challenges, a goal to develop a formal governance framework was agreed upon by the main apparel conglomerates and government representatives under the UN Sustainable Development Goals in a series of UN initiatives launched since 2018 (Gardetti & Muthu, 2020). Until that moment, governance of the apparel sector had never been designed or applied with sustainable production goals. It had been focused on issues of free trade, market access, and more recently, labor protection (Anguelov, 2015; Anner, 2009, 2011, 2019). As already explained, the urgency to undergo legislative innovation to address environmental impact, as well as labor issues, stems from the significant ecological impact of the industry, as it continues to grow (Pal & Gander, 2018).

With the on-going retail innovations, including fast fashion, super-fast fashion, and online commerce, sales volumes keep on increasing. Having followed a steady trajectory of expansion, the retail value of the clothing and textile business is estimated at $1,500 billion annually, with future growth rate projected to increase in terms of both volume and value. According to the International Cotton Advisory Committee (ICAC), annual textile consumption is projected to grow at a rate of 3.1 % until 2025 (Sodhi, 2017). Such expansion is most-directly linked to the on-going expansion of fast fashion retail (Nucamendi-Guillén, Moreno, & Mendoza, 2018; Wen, Choi, & Chung, 2019).

Along with fast fashion, two fairly new business models that are transforming retail are expected to play significant role in the future growth of the industry. They are "mass customization" and "athleisure." Mass customization is offering product lines with choice, where customers can pick apparel features or even, as is the case with American-based *Stitch Fix*, alter and/or mend old clothes (Choi, 2013; Fiore, Lee, & Kunz, 2004). Athleisure is sporty apparel, including clothes, shoes, and outerwear, that people wear regardless of a physically active lifestyle and almost never to actually exercise (Craik, 2019; Lipson, Stewart, & Griffiths, 2020). These two platforms merit analysis as they have the potential to redefine production needs in materials, purchasing behavior of consumers, as well as the emerging sustainability governance of the industry.

Mass customization developed as a niche platform for the better part of the last two decades as technology allowed more customer input into the production process. First introduced in the late 1980s, mass customization was more or less an exploratory platform for high-end design (Rahman & Gong, 2016). At its core, it is a modern brand-based seamstress model, meaning making tailored clothes based on customer specifications. It was not until the beginning of the 2000s, however, that mass customization became a viable mass-market commercial model. Even then, it remained mostly vested in luxury and prestige brands that allowed customers options, in what is called "co-design," where customers can choose features from a variety of available options (Azuma & Fernie, 2003; Ulrich, Anderson-Connell, & Wu, 2003).

About a decade ago fashion scholars and experts on consumer marketing and relations made the link between mass customization and "sustainability," exploring the opportunities to reduce waste (Black & Eckert, 2010; Niinimäki & Hassi, 2011). The assumption is that more input into the design process, better measurement information, and a responsive system of customer relations based on smart technology, would allow for the direct production of apparel to exact customer specifications. Hence, less "needless" variety would have to be manufactured, leading to lesser degree of discarding of unwanted clothes. Under such assumptions, mass customization can lend itself to claims of sustainability, as Black and Eckert (2010) explain, but only in terms of reducing waste.

In an effort to develop business models that embrace the waste-reducing quest, under mass customization specifications, certain innovative fashion retail start-ups have developed the emerging commercial platform of "rent" vs. "buy" (Hu et al., 2014). Todeschini et al. (2017) provide the most recent and comprehensive analysis of sustainability features in apparel retail innovation, including mass customization under the "born sustainable" classification. The authors track the growth of "rent" vs. "buy" commerce with a movement to create a "fashion library" culture. Companies that rent outfits include the appropriately named *Rent*, *Runway*, and *LENA*.

Interest in "rent" vs. "buy" is growing, proliferating outside innovative start-ups to the global conglomerates (Hanbury, 2019). Prestige brands, such as *New York & Company* and *Bloomingdales*, are offering renting monthly subscription-based services. Even "masstige" brands, such as *Banana Republic* and *GAP* offer rental options, leading *H&M* – the powerhouse of fast fashion retail – to announce plans to launch rental services in its flagship Stockholm store for premium-priced lines, focused on products that include recycled inputs (Dowsett & Fares, 2019). A noble plan, or another PR opportunity for *H&M* to showcase commitment with vacuous promises? Time will tell.

Rental business model options are currently in the developmental stages of proliferation and still at relatively high price points. It makes sense to rent expensive garments. The challenge is launching customization retail, be it with rental options or custom-made options, in the broader "mass" fashion market. A successful example comes from Todeschini et al. (2017: 768) who describe a collaboration of Italian retailers to launch, as the authors put it: "an Italian e-commerce platform." It offers entirely made-in-Italy men's clothing that is (a) made to measure with a 3-D configurator, (b) locally sourced, (c) has a face-to-face option for tailoring in major cities, and (d) offers up to 10 million combination options of choice. Collaborations as these can lead to customization retail "at scale," as is the industry term, meaning able to serve millions of customers globally. The mass-customization retailers that are tackling the challenge of "production to scale," at masstige to prestige price points, are US-based *Stitch Fix*, *Trunk Club*, Germany's *Adidas* and start-ups such as *Suit Supply* and *True & Co*, and Delhi-based *Pernia's Pop-Up Shop* (Sodhi, 2017, 2018).

Along with such hopeful innovation in product development and retail models that explore commercial ways to reduce the industry's carbon footprint, there is also a style redefinition in what consumers value as fashion. This redefinition is a growing preference for, what Chapter 1 in this book described as, "activewear" and "sportswear" in traditional fashion retail. The new metamorphosis of this decades-old-by-now, and still growing, cultural embrace of athletic rather than fashionable apparel is in modern-day's growing popularity of athleisure. Athleisure blends well with fast fashion as it is mainly a design-based innovation. It can be mass-produced and positioned with the same or similar turnover rates as fast fashion product lines. However, it makes no claim to "fashion," in terms of being reflective of high-fashion trends.

The volume of athleisure commerce keeps increasing. According to Morgan Stanley research, its growth rates were strongest between 2008 and 2015 (Sodhi, 2017). In 2014, CNBC business analysts event went as far as claiming that the athleisure trend "spells the death of denim" (Korber & Reagan, 2014). An example of how important the trend is for the whole industry comes from the opening lyrics of pop star Cadri B's 2019 hit "I like It," where she sings: "I like those Balenciagas; the ones that look like socks." "Those Balenciagas" are form-fitting boot/sneaker hybrid that one pulls on their feet like socks. *Balenciaga* – one of the oldest luxury fashion houses – invests in the creation of futuristic-looking athleisure products, targeting a customer base that defines style from popular culture cues. Along, of course, with jewel encrusted stilettos.

Balenciaga has no choice. In order to survive in a global marketplace defined by fast fashion and social media, where style is demarcated by influencers outside of the circles of design royalty, all retailers have to offer product that can blend style with comfort and performance. The Balenciagas of the industry must learn to survive not only against the fast fashion Zaras and H&Ms, which increasingly, and more baldly, diversity into luxury product lines, as their social importance rises. They must also survive against the industry's tech innovators, such as Boston-based *Ministry of Supply*, a start-up launched by two MIT-trained textile mechanics engineers in 2012. *Ministry of Supply* defines its mission as "to design and construct garments true to the form of human body… where form and function intersect," boasting the use of technologies used in NASA space-suit exploration gear. The promotional materials on *Ministry of Supply's* home page include quotes from *Vogue* magazine, naming it "one of the 59 digitally native brands you'll see everywhere in 2019." I still haven't at the end of 2020, but the point is that with such innovators, the industry is now entering a new phase – "engineered apparel" that is "digitally native."

Engineered apparel is the launch of clothes made from "functional fabrics," traditionally developed for athletes and professionals in fields that needed protective clothing (Hayes & Venkatraman, 2016; Shishoo, 2015). Digitally native retailers, such as *Ministry of Supply*, are well, exactly that, also referred to as "online only" or "online first" retailers (Bell et al., 2018). The term "digitally native" is most-often used to describe the consumer behavior of millennial and Z generations, as having grown-up with technology in a cultural cyber space (Howe & Teufel, 2014). Members of those generations favor online communications, advertising, and commerce, feeling comfortable with "online only" retailers. The problem is that these "functional textiles," driving "digitally native" "athleisure" product development, are synthetic.

Innovation in material development had been decoupled from sustainability concerns. Even as fast fashion commerce was escalating and concerns were being raised of its ecological impact, fashion designers continued to create and laud the use of "modern fabrics." In the documentary *DRIES*, Belgian designer Dries van Noten tracks his most important collections that define his 30-year career to establish him as one of today's most innovative designers. He explains that it was his 2007 summer collection that defines him as "pushing things really forward" and he means using materials such as taffeta and silk and polyester blends to make fabrics look different and move differently (Holzemer, 2017). The designer explains that from that point on, his designs were about contemporary clothes and contemporary fabrics. The issue is that those "contemporary" fabrics and their production is the core problem in textile toxicity. It is these types of fabrics that, as explained in Chapters 3 and 4, constitute the main problem for fashion circularity – they cannot be recycled. They are toxic in production and care and they are ubiquitous in almost all clothing items. Today one would be hard pressed to find a garment that is, as used to be known,

100% cotton or other mono-yarn material. The embrace and promotion of "contemporary fabrics" by designers is the core issue that frustrates the fashion sustainability movement.

There has been a decided move away from the interest in "natural" fibers, best captured by the "organic cotton" movement of a decade ago, into innovation in athleisure production that is focused on "functional fabrics" (Tadesse et al., 2019; Yan et al., 2018). Functional fabrics define apparel innovations which include:

(a) odor-controlling technology (Klepp et al., 2016; McQueen & Vaezafshar, 2019), most successfully commercialized by South Korean conglomerate *Polygiene*
(b) non-irritating graphene textiles (Malhotra & Mandal, 2019), commercialized by Italian *Directa Plus*
(c) metal-organic framework (MOF) powders (Rose et al., 2011), developed to remove toxic compounds in "protective clothing," for military and first responder professionals

Product features in such developments promote lower "toxicity" and the use of recycled inputs from reclaimed polyester. As discussed in previous chapters, however, such reclaiming is far from constituting a viable carbon-footprint-reducing, "circular" fashion economy. Developing fabrics that can be promoted for their sustainability-improving features is an on-going process, which currently, has not yet been able to deliver products that are mass-market ready. Yet, innovators are emerging, such as the Swedish *Re:newcell* that is successfully pioneering denim recycling at scale.

The company's operations are featured in the special segment *The New High-tech Way to Recycle Clothes* by BBC's series *Click*, which tracks global innovation trends across industries (BBC, 2019). *Re:newcell* uses, as is the emerging term "climate positive" operations to turn reclaimed denim into viscose. It relies on pre-sorting suppliers that collect and disaggregate the denim to remove the stitching, which is not made of recyclable natural fibers, and then an "eco-friendly," according to the company's promotions, "chemical process" is used to break down the fabric and de-dye it.

Although not specified in the program or on *Re:newcell's* media platforms, these processes most likely involve the use of enzymes. To remove indigo-based dye, peroxidase enzymes or laccase enzymes are used. For Sulfur-based dies, esterase enzymes can be used (Singh, Singh, & Singh, 2015). The use of enzymes as alternatives to chemical auxiliaries in textile coloration (the term used for both dyeing and removing color) has gained popularity in research for its relatively lower environmental impact (Bansal & Kanwar, 2013; Fu et al., 2012). How much lower? It is still being evaluated and tested by cleaner production textile scientists.

Not all color can be removed completely in such processes. Any remaining color has to be removed through "chlorine-free" bleaching, as *Re:newcell* claims. Then the wet pulp is dried up into sheets of thick canvas, resembling that used in oil painting, called "circulose" to be sold on to the next tier of companies that use this canvas to turn it into thread. That thread is not yarn; it looks and feels like natural cotton wool. The circulose is then sold on the yarn weavers. Although the film refers to the product as circulose, it is in effect a type of lyocell or viscose, which are made from wood (Sealey et al., 2004; Zhang et al., 2018) and have been part of niche eco-textile development for a few decades.

Yarns made from lyocell and viscose are fragile and have very limited commercial use, therefore, they have not gained mass-market popularity. This fact is not mentioned in the documentary *The New High-tech Way to Recycle Clothes*. The segment ends on a positive note, promoting *Re:newcell's* product – circulouse.

As with previous such programs, the narrative ends with an upbeat message, promoting a product. The problem is that this product is just one of the inputs tier 3 suppliers can choose for weaving fabric. How the choice of a textile factory to use circulouse helps with the most toxic operational steps in fabric production, outlined previously in Chapter 3, weaving and dyeing of yarns, is not discussed. The next frame in the film shows the narrator handling a bright yellow casual dress, stating: "…it's pretty nice; it's made of viscose and it's recyclable again…" But is it? With existing chemical recycling options that would turn it back into circulose? Who is going to recycle it? How fragile would the reclaimed pulp be? The narrator claims that the yellow dress can be "recycled up to five or six times." She then gives an astonished smile, and the scene ends.

Explaining the market dynamics, the CEO of *Re:newcell* Mattias Jonsson states that the company opened at-scale operations in 2018. It prototyped the "yellow dress" during "fashion weeks" in 2014 to showcase designers what is possible to create with circulose. Mr. Jonsson states that the capacity of the production facility is up to 7,000 tons annually, which equals the weight of 30 million t-shirts. He claims that *Re:newcell* is the first company in the world to be producing on industrial scale. The camera work shows the evidence – sophisticated heavy machinery in a large industrial complex (of course on the bank of a river) in Sweden. It must be noted that this example comes from the nation that gave the world fast fashion, where its economy is dependent on the success of *H&M's* global operations, and where the government has implemented laws to incentivize such operations. Paras et al. (2018) explain that the Swedish tax agency has "recently" created a value-added exemption for organizations that have collection and/or processing operations for second-hand clothes. *Re:newcell* must be benefiting from such a tax exemption. To relate such support to the concept of nation branding, covered in Chapter 2, the Swedish government is actively promoting its sustainability initiatives, while at the same time, engaging in the international promotion of the Swedish firms (Mansson, 2016). That is why, consumers the world over are eager to believe *H&M's* social justice proclamations and its extremely effective greenwashing campaigns.

Even with such government support, there is not enough resources to keep operations profitable for *Re:newcell.* The firm relies on denim, imported mainly from the United States, and at this point, treating denim is the only operation the company can do. The reason is that, denim is comparatively easy to de-dye, but only the denim that has been dyed with 100% indigo-derived dye. However, most denim has been dyed with a combination of indigo and Sulfur or 100% Sulfur-made dye, which is the case for black and colored denim. Those types of dyes are not as easy to remove. Textile engineers are experimenting with innovations (Buscio, Crespi, & Gutiérrez-Bouzán, 2015; Maryan, Montazer, & Damerchely, 2015; Silva et al., 2018), including the use of lasers (Dascalu et al., 2000), yet from the literature examined in this book and its prequel, there is little evidence that such technologies are being deployed on a commercial scale.

The main reason is simple. The deployment of such innovations at scale would be very expensive. A lot more firms similar to *Re:newcell* would be needed to generate the demand for a separate tier of engineering equipment firms to respond by producing the necessary machinery.

There are no other materials that *Re:newcell* can process, which means that this promise that the "yellow dress," prominently displayed beyond the CEO of the company during the interview, would have to be "recycled up to five or six times," not there. Then where? No such questions are asked. The segment ends with the typical greenwashing promotional statements that *H&M*, an investor in the company – a piece of information strategically saved for the last minutes of the segment perhaps to increase the "feel good factor" of the story – will "soon have clothing originated from this process on their shop floors." Yet, as expected, no discussion of costs accompanies this segment, or the other related segments in innovation in the *Click* series dedicated to sustainable fashion.

So far *Click* has produced few episodes on innovation toward circular transformation. In the above-summarized segment *The New High-tech Way to Recycle Clothes*, the ending minutes even offer the claim that the industry's goal is to be fully circular or sustainable by 2030. We, as the audience, want to hear such news. And we, as the industry and policy professionals, want to have such goals. The problem is that reaching them requires time and resources. So far, if history has taught us anything about innovation in the industry, it is that it has taken over 20 years of political and social action to get the first and only industrial scale production facility open that treats the basic fiber in clothes – cotton – and only one type of clothing item made from it – denim. Making claims and promises that all fabric, including poly-blends, synthetics, and leathers, would be able to be included in operations that would transform the whole industry to "fully circular" is unwise, if not irresponsible.

The main issue with such promises is the fact that innovations in material improvements are just beginning to gain the interest of fabric manufacturers. Furthermore, depending on market niches and opportunities, producers can focus on different features that can be branded as "sustainable" innovations. For example, attention is placed on the integration of "functional powders" into fabrics (Yadav et al., 2006; Yang et al., 2012) – a technology refined by French *Fibroline* with the use of high voltage generators to alter electric fields. This is a process that does not require water or solvents.

The goal to limit water use is also reflected in innovation for "100% water-free garments," (Samanta, Basak, & Chattopadhyay, 2017; Pal, Chatterjee, & Sharma, 2017). The best example to date is the much-publicized *Phoenix Jacket*, launched in 2019 by outdoor brand *Marmot*. The promotion claims that it is the first-ever garment to be made without the use of water in any stage of production, manufactured with technology employing solution-dyed yarns and dry fabric finishing. The problem is that it is all nylon – a petrochemical product, the production of which is akin to a plastic bag – and it retails at luxury price points. *Marmot's* website lists the cheapest option at $175 for what in essence is a light raincoat. How many customers will be concerned enough about their ecological footprint, wealthy enough, digitally savvy enough, and most-importantly, be convinced that the *Phoenix Jacket* is indeed a "sustainable" garment to buy it? How many will question it and opt for a light rain coat from *Target*, one fifth the price? Time will tell.

With such noted limitations in clarity, the digitally native, modern apparel consumer exists in a fashion industrial culture permeated with messages of sustainability innovation. Brands are using the concept for their own promotional purposes, boasting self-established norms and regulations to showcase commitment to environmental stewardship. What started a decade ago as a niche in the industry, at the time referred to as "eco fashion," has now become a constant promotional rhetoric for retailers and their suppliers, who feel the pressure to show evidence of sustainability in their operations. The quest to incentivize viable improvements toward quantifying such evidence has moved toward redefining the role of governments in the fashion market.

Formal Governance: Foundations and Evolution

Today most industry-level regulation is focused on labor. Yet, despite effort to address sweatshop working conditions, establish health and occupational safety guidelines, and regulate enforcement to protect workers, problems continue. For example, despite all the social attention and outcry for the poor working conditions in facilities in Dhaka, Bangladesh after the tragic collapse of the Rana Plaza complex in 2013, in 2020 173 factories were found unsafe, according to standards of the Bangladesh Department of Inspection for Factories and Establishments (DIFE) (Glover, 2020). The main reason is regulatory capture. Industrialists and politicians are part of the same social elites and are often both.

Politicians have industrial interests and with those, have direct monetary incentives to increase profitability of their ventures and keep operating costs competitively low. In the developing nations most dependent on apparel production, pressures rise congruent with development. As development levels increase, so do labor costs as wages go up (Cui & Lu, 2018; Jong-Wha & Wie, 2017; Yang, Chen, & Monarch, 2010). This is very much the case in China today, as it was in South Korea in the 1980s when textile producers started locating out of South Korea into lower cost locations (ESCAP, 2008; Maurice & Hermann, 2017).

Indirect costs of waste and pollution mitigation also rise with institutional development, as the general public values cleaner environment and laws are developed to curb pollution. As already explained, this process is the focus on academic works on the Kuztnets Curves from across industrial sectors (Antonakakis & Collins, 2018; Piketty, 2006; Rudra & Chattopadhyay, 2018). The most direct effect of such legislation is to increase the average operating costs of producers by making them pay for the deployment of purification and filtration technologies (Cherniwchan, Copeland, & Taylor, 2017).

In terms of legislating incentives for such internalization of the high social costs of fashion pollution, since the 1990s Europe has been the leader in toxicity mitigation laws, banning the use of certain inputs, such as the highly toxic azo dyes (Brüschweiler et al., 2014). Azo dyes do not biodegrade and when expelled into run-offs, tend to "bioaccumulate" in aquafers, having a negative effect on safe farming and of course, wild life (Bafana, Devi, & Chakrabarti, 2011). Since most dyeing and finishing happens internationally, the current legal efforts in Europe, established under the REACH regulation of 2007, differ from the set of 1990s laws, including Germany's MST and MUT laws that set standards for pollution contents in finished goods and production processes at the national level (Anguelov, 2015: 115–116). MST and MUT applied only to Germany. REACH stands for Registration, Evaluation, Authorization, and Restriction of Chemicals, and is under the scope of the European Chemical Agency, with authority (at least on paper) over all EU member states.

The earlier set of laws banned the commerce of product made with certain toxic chemicals, mainly colorants. In general, they applied to imports into one specific nation, as in not allowing clothes made outside of Germany with banned dyes (in relation to the MST and MUT laws) to be sold in Germany, for example. REACH, on the other hand, applies in production process regulations for the whole of the European Union and across operational tiers. REACH is supposed to regulate the manufacture, import, marketing, and end-use of chemicals.

REACH's authority is broad, and as such, prone to high level of administrative discretion. As its acronym describes, it does indeed "Register" and "Authorize" chemical use. Yet, according to the language on its website, it only "Restricts" it when its "Evaluations" reveal that the use of certain chemicals, at certain levels, poses "risks that cannot be managed."[1] With respect to apparel production Jacometti (2019) explains that REACH provisions can be applied to chemical companies, textile manufacturers, and leather tanneries in the use of colorants, auxiliaries, and biocidal additives, which are used in leather and textile treatment as fungicides to inhibit bacterial growth. Since 2007, these provisions have guided production inside the European Union. However, since most apparel inputs are actually manufactured outside of its jurisdiction, new legislation is being developed with respect to global supply chains.

The new legislative efforts, set forth by *Resolution of 27 April 2017 on the EU Flagship Initiative on the Garment Sector* 2016/2140, put the attention on supply chain transparency (European Parliament, 2017). It merits to describe this Resolution, as merely an effort because its language calls for the "development of a legal framework" that includes "measures on due diligence obligations" (Jacometti, 2019: 27). How and under what jurisdiction this "framework" would operate, under what mandates, and most important, with what oversight and enforceability of compliance, is unclear. Afterall, as Niinimäki et al. (2020) show, over 80% of clothes sold in the European Union are not manufactured in Europe. If anything, it is troublesome that the precedent, which the Resolution seems to propose to follow, is based on voluntary participation of EU Ecolabel.

EU Ecolabel was launched in 2011. Jacometti (2019) offers a detailed breakdown of its legal provisions and explains all are voluntary, there are no punitive measures for non-compliance, and all brandish bombastic, ambiguous language, such as "circularity" and "product and organizational footprint." The caveat is the EU Ecolabel platform operates like any other of the "eco-fashion" voluntary certification bodies, leading among then being OEKO-TEX, and until recently *Made-By*, proven to be irrelevant and even defunct. *Made-By* filed for bankruptcy in 2019.

These bodies are, in effect, consultancies that one can cavalierly state "help" brands make "better choices." The reality is that they just allow brands to create reasons for price mark-ups of certain product lines, without holding them accountable to any commitments in their overall production practices. Writing for *Eluxe Magazine*, Caric (2019) ranks the 10 leading "ethical" fashion certification bodies and offers honest criticism in each one's special provisions of the notable limitations for oversight of compliance. The conclusions do not show any convincing beneficial evidence.

One goal is emerging from the analysis of the formal legal actions taken in Europe. It is the reduction of waste (Jacometti, 2019; Moorhouse & Moorhouse, 2017; To et al., 2019). The Waste Framework Directive 2008/98/EC of the European Parliament "establishes some fundamental principles," as Jacometti (2019: 27) puts it. The three main principles are as follows:

1) obligation to handle waste in best effort to protect public health
2) the principle of waste hierarchy and
3) following "the polluter pays" principle

These principles are broad, and in that fact lies the problem. Specifically, principles (1) and (2) can arguably be described as symbolically vacuous. Whose "obligation" and "best effort" to protect public health does the first principle impact? Federal or local government, waste management authority, out-contracted waste management providers, or producers? How are infractions to this principle quantified, and how are infractions to be adjudicated?

The second principal of "waste hierarchy" is even more problematic. It is a step-wise goal in waste management legislation to consider options in mitigating environmental damage before formal waste disposal at the processing level. The hierarchy should follow this order: (1) prevention, (2) preparing for re-use, (3) recycling, (4) "other" recovery, and (5) disposal (Giacometti, 2019). It is unclear who has oversight in each of the links in the order, or how compliance is to be measured, with what level of discretion, and under what jurisdictional authority (Corvellec, 2016; Gharfalkar et al., 2015).

In textile waste management, (2) and (3) are of specific importance because (2) can serve as a legal incentive structure for business innovators such as *Stitch Fix* to employ discarded fabric and trim. However, (3) is unfeasible because, as explained in Chapters 3 and 4 of this book, it is almost impossible to recycle fabric in a way that can be seen as a step toward protecting the environment. Mandating "recycling" in waste management is currently akin to mandating a disposal method for discarded apparel that separates it from other waste. That is it. Yet, how specifically that legal platform is to be implemented is unclear.

Are clothes to be included in recycling bins with paper, plastic and glass? Are they to be collected separately, which currently is the standard platform? This standard platform creates issues because citizens must first make the choice to recycle their old clothes and then have to bear the transaction costs of actually locating and making a trip to an appropriate drop off point. Research has shown that very little of this dynamic is happening (Ekström, & Salomonson, 2014; Kapoor & Khare, 2019; Sandvik & Stubbs, 2019).

As already explained in previous chapters, only apparel made from either purely natural yarn fabrics – cotton, linen, wool, or purely man-made, and only certain types of polyester – can be recycled, and not in ways that are ecologically safe. As businesses have developed around both, more social attention has been placed on treating natural fibers, which is understandable due to the general sustainability mentality to discourage the use of petrochemical products, which include polyester. Yet, there is an important dynamic that must be incorporated in policy and legal design to incentivize circular fashion, and that is the fact that apparel can be made much more cheaply from polyester inputs, which can negatively impact recycling efforts of natural yarns. Kapoor and Khare (2019) offer the most compelling example of this unfortunate reality analyzing a cluster of factories in India that, for a decade, had specialized in recycling wool and using it to manufacture blankets for emergency relief and disaster first-responder use. First, one must note the end product of such recycling. It is not clothes retailed in stores that must be durable and easy to care for, withstand repeated washing or dry cleaning, come in different colors and textures, and most importantly, appear enticing next to unrecycled substitution options. The end product is blankets, and not for home-product commerce, which would require them to have features similar to clothes, but for disaster relief. This fact, although not discussed by Kapoor and Khare (2019), captures the reality of what can be manufactured from recycled yarns – product that is fragile, lacks versatility, and therefore can be of limited commercial use. The authors offer sobering input of the fact that cheaper, finer, lighter, and softer polyester blankets from China are driving the sector out of business. The Indian recycled wool blankets wholesale at around $7 US – the price reflects the complicated import structure of discarded apparel mainly from the West. The Chinese competitors come in at $2 a piece. Since the customers are government organizations and relief military units without clear sustainability mandates, their procurement officers have no incentive to choose the more expensive products and they do not.

Such reports show a developing interest in the academy to understand fashion economic incentives when it comes to sustainability goals in a reality of substitution options that are more competitive because they are cheaper. Such works also honestly explain the production actuality in apparel, which is concentrated in few under-developed nations that are heavily reliant in their national economies on textile exports. Production for the European market happens there – Bangladesh, India, Pakistan, China, Viet-Nam, Cambodia, Ethiopia, North Korea and about 10 others (Miroux & Sauvant, 2005). What are the implications of the European Commission Waster Framework Directive 2008/98/EC for processes that occur in those nations? None, according to the very detailed analysis of the legal provisions of the Directive and in particular its Articles 5 and 6 that have been rewritten to denote waste to be valued as a resource (Gacometti, 2019). With respect to regulating chemical use, Niinimäki et al. (2020) show that because the majority of textiles imported into the EU are "partially treated" items, which are then "finished locally," the current oversight structures make it difficult to understand total chemical usage.

The language of the European Commission Waste Framework Directive makes it clear that all provisions apply only to EU member states. The wording of compliance guidelines denotes a lack of enforceability with phrases such as "the Commission should be empowered to adopt implementation acts." It should be, but it is not. The Commission does not set any criteria at the European Union level, leaving the establishment of criteria to member states.

This platform has two major loopholes that allow for the business-as-usual dynamics of the industry to continue with very little impact or real sustainability improvements. The main one is that none of the provisions have jurisdiction over international suppliers or a legal way to incentivize imports from any sourcing markets that may be willing to adopt and comply with cleaner production policies. The second loophole is that the directives are to be interpreted, internalized, and set into policy by individual member states.

Such lack of clarity on how directives apply through the global supply chain that manufactures clothes for the European Union is behind the major challenge for the third principle of the European Commission Waste Framework Directive – "the polluter pays" principal, which has guided European waste law for the past several decades (Van Calster, 2015). In effect, all the directives behind the principal apply to producers. Yet, there is an extremely small portion of any apparel production that actually happens in Europe. Again, most of it is in the processing of semi-finished garments, not in the fairly more toxic stages of thread and fabric weaving (Niinimäki et al., 2020). Although some facilities are operating in Eastern Europe, there is very little evidence that they are complying with such EU principals (Anguelov, 2015: 100-102).

How can "the polluter pays" principle impact the international suppliers of apparel production components, as tracked in Chapter 3, is still a subject to debate. Some progress has been made with the advent of "environmental scorecards" (Garcia-Torres, Rey-Garcia, & Albareda-Vivo, 2017; Madsen & Slåtten, 2013; Turker & Altuntas, 2014). France is currently exploring legislative options to mandate putting environmental scores on labels (Remington, 2020).

The problem is that even if such legislation is passed, a system has been developed in the past decade around scorecards that is, in effect, a separate industry of consultancies. As with any industrial free market structure, there's competition. *Made-by* filed for bankruptcy while its competitors thrive, including San Francisco-based *Sustainable Apparel Coalition*, developers of *Higg Index*, to be analyzed in some detail shortly. The main competitive dynamic among such consultancies is to *disagree* with each other on prioritizing operations, foci, options, and even chemical inputs.

Such firms offer brands their services in analyzing production and operations steps, measuring the ecological impact at each, and developing a guide for improvements. The specific deliverables are most-often advising individual suppliers in a brand's production chain to choose less-toxic chemicals, when feasible. Currently, one such consultancy, Amsterdam-based *Go-Blu* is developing an app that would track chemical inputs for textile factory managers. The issue is that less-toxic options are not always available for specific color combinations and products. As dyeing is the most toxic process, the often-given advice is to opt for "natural" dyes, which bring their own set of problems.

One must bear in mind that before they are used commercially, natural dyes are produced in very water, resource, and labor-intensive ways from plants, seeds, lichens, fruits, and seeds. Their manufacturing has a large ecological footprint. Natural dyes are of a lesser color intensity and fade in natural sunlight, from body heat, and in washing. Therefore, garments made with natural colors would be discarded fairly fast by consumers – not necessarily improving the fast fashion consumption problem. But most importantly, they require the use of toxic fixatives called "mordants" to bond to the fibers, which are the real culprit in both natural and synthetic dyes (Prabhu & Bhute, 2012).

Mordants containing metallic salts are most-often used to improve the vividness in natural dye coloring. Those compounds – Potassium Dichromate (chrome), Stannous Chloride (tin), Copper Sulfate and Iron Sulfate – are lethal in industrial concentrations (Ransom, 2020). Furthermore, natural dyes can only be applied on natural yarns. Polyester and poly-blend fabric producers cannot use them. Therefore, brands that rely on athleisure product lines, athletic brands, and most fast fashion brands that offer poly-blend clothing, which let's be honest, is most often the case, cannot honestly be expected to improve their environmental scorecards by asking their dyers to opt for natural dyes.

The platform of scorecarding is not only voluntary, it is also non-binding in adherence. Brands can choose to follow the recommendation of the consultants or not. Most importantly, they can choose to either rely on consultant certification services (or not) in their own environmental self-assessments. Those are the reasons why scorecarding is an industry of consultants. Most of what they do is try and "convince" brands to "sign up" for their services. The problem is that participation raises the operational costs for brands on two fronts. One is direct – paying the consultants. The other is a more diffuse spike in operational costs that has to cover the entire supply chain. It means implementing standards developed by the consultants on use of inputs and also developing a compliance and monitoring system with suppliers. Despite such challenges, there is evidence that a platform of voluntary self-regulation initiatives is emerging.

Voluntary Governance in Sustainability Compliance: The Implications of Self-Regulation

Today, the *Higg Index*, launched by the *Sustainable Apparel Coalition*, is among the most-celebrated, voluntary, self-regulation collaborative initiative in the sector. Industry insiders, as well as emerging academic analysis, are linking it to the UN Sustainable Development Goals as the retail module to follow (Gardetti, 2015). The Index is being used by hundreds of textile and footwear manufacturers, brands, retailers, and other stakeholders (Sodhi, 2018). It consists of a "self-assessment" suite of tools.

The suite of tools is comprised of three modules that use a standardized scoring methodology to rate the performance of an apparel company's brand, facilities, and products. One of the issues with the current state of the Index is the lack of connectivity between the product and facility modules (Connolly, 2015). It is due to the fact that "facilities," in terms of factories and production and processing centers under brand ownership, do not exist in the "born global" retailers of today. *H&M* is one such conglomerate that does not own any factories (Wada, 1992), but relies on independent producers through a series of tiers, as explained in Chapter 3. The reality is that Tier 1 suppliers, often the sewing facilities, may be the extent of brands' knowledge of who is in their supply chain. When it comes to order placement, it is Tier 1 purchasing offices that source inputs from Tiers 2 and 3. Brands rarely have knowledge, much less oversight, of Tier 2 and 3 firms.

Although brands have different ways of defining "tiers," and everything shifts when there are agents involved that help broker deals between brands and manufacturers, none of the tier suppliers' clients have supervisory functions or powers. There are initiatives underway to address this disconnect, and some research is examining their effectiveness. For example, M. Tachizawa and Yew Wong (2014) explain the evolution of sustainability mandates through the tier system and note that often first-tier suppliers train managers in lower tiers to use environmental databases. It is a challenging undertaking because traceability through the multiple links in Tiers 2 and 3 is not executed in an integrated supply-chain format. That is the case because ownership of tiered facilities is under multinational conglomerate structures of management bodies, called "groups," which vary greatly in nationality, in terms of headquarter location and incorporation (de Abreu et al., 2012; Narwal & Jindal, 2015; Singleton, 2013).

Arora et al. (2004) use the phrase "stateless corporations" to denote that these conglomerates own a variety of firms located in different countries, incorporated often in tax-havens, funded by international banks and wealth funds. Whose laws do they abide by? On paper, should there be an issue in international dispute litigation, the corporate parent owner and the laws of the nation of incorporation of that entity would apply. Yet in fashion, when it comes to "compliance" with "norms, guidelines, and standards," which are legal provisions without binding mandates, infraction can only be handled by a choice of customer to withhold business. For example, the *Shinest Group* is one of *H&M's* "preferred suppliers" and its facilities in Bangladesh produce *H&M's Conscious Collection*. The *Shinest Group* consists of 12 different factories, specializing in different processes – from assembly to embroidery to manufacturing trim, meaning buttons, zippers, and toggles – and producing for many different clients. To put the *Higg Index* into context here, the Index's mandates apply to *H&M*. Should *H&M* choose to follow its self-assessment suit of tools, it bears the responsibility to monitor the operations of the *Shinest Group* for compliance. Investigative analysis of this hypothetical dynamic shows that not only not to be the case, but also exposes alarming exploitation and greenwashing under this seemingly well-established partnership to manufacture a product line that boasts to be the flagship sustainability initiative of *H&M* (Maurice & Hermann, 2017).

The latest version of the *Higg Index 2.0*, launched in 2013, aims to develop a standardization for production organizations that include the brands and their tiered suppliers, on how to measure and evaluate environmental performance. Blockchain technology can be implemented and much experimentation is on-going with the development of "circular digital ID" for apparel. It is unclear yet however, exactly how the measurement of such environmental performance translates onto labels.

Analyzing the specific outcomes of the *Higg Index*, Nidumolu et al. (2014) posit the *Higg Index* is influencing capital investment, as in providing a platform for funding sustainable innovations in material science and operational performance. The authors go as far as claiming that the Index is changing operational behavior. It is linked to the advent of formal policies governments are adopting, such as Zero Liquid Discharge, Zero Waste to Landfill Facilities and Zero Discharge of Hazardous Chemicals (ZDHC) in India, Pakistan and Bangladesh (Rajamani, 2016; Yaqub & Lee, 2019; Tong

& Elimelech, 2016). In essence, those policies mandate that factories have a treatment facility. Unfortunately, as explained in Chapter 3, these policies are not followed stringently by factories and emerging effluent treatment innovations are not widely used.

With such evidence coming out of on-site compliance investigations, it is not surprising that the latest academic analysis of the effectiveness of the *Higg Index* shows it to be limited, at best. University of California, Berkeley's Professor of Environmental and Labor Policy Dara O'Rourke, who since the 1990s has been among the leading environmental policy scholars focused on apparel, with PhD candidate Niklas Lollo, offered a comprehensive report on *Higg Index* compliance, based on three years of data from leading facilities in Bangladesh and China (Lollo & O'Rourke, 2020). The report analyzes the Higg Facility Environmental Module (FEM). FEM is just one component of the *Higg Index* suite of tools. Thousands of factories claim to use it, in what is becoming the standard of industrial environmental monitoring – self-assessment. In effect, each facility uses FEM to track its chemical use, energy use, and labor and safety policies. Oversight is internal, meaning handled by factory management, and the information is used for building customer relations. Factories try to show evidence of sustainable practices to appeal to clients. Hence, Lollo and O'Rourke (2020) sum up in their findings that factories share this information privately and with discretion between suppliers and buyers only. Greater transparency and sustained compliance information is unavailable. That is why, on-site researchers on ETP and LZD use find major lack of veracity in compliance claims (Anas, 2015; Dasgupta et al., 2015; Holkar et al., 2016; Mohan et al., 2017; Sakamoto et al., 2019).

A system has formed of mutually beneficial collusion. It is because both suppliers and buyers have an incentive to keep costs as low as possible. As buyers specifically evaluate suppliers mainly on lowest costs, they have no incentive to raise concerns of non-compliance in this system where information is shared from supplier to buyer. Putting buyers into the overseer role only increases the perverse incentives to maintain a cost-cutting competitive structure.

Note

1 See: Understanding REACH, available at: https://echa.europa.eu/regulations/reach/understanding-reach

Chapter 7

The Way Ahead

Philosophy and social sciences professor Chrysostomos Mantzavinos (2004) logically explains that in order for market change to occur, a cultural change must precede it, and culture is slow to change. It is clear from the tomes of research on the ecological problems of apparel economics that overconsumption and discarding of cheap clothes are the market problems that should be the subject of change. It is a market problem, because the negative externalities created are not being sufficiently addressed by price to pay for the high social costs of environmental degradation. Social costs are caused by specific market actors – in this case producers involved in toxic manufacturing – however, they are born by society at large. The consumers of toxic products pay a price that only covers production costs. The social costs of public health damage, loss of bio-diversity, and climate change are impossible to internalize with a market structure, because they are intertemporal and diffuse.

To address such market inefficiency, as explained by the works on Environmental Kuznets Curves in Chapter 5, governments enter market activity to impose regulation. In the fashion sector, it is this process that is beginning its impetus. Formal governance calls for action were put forth in 2020 by the United Nations Sustainable Development Goals for Better Fashion. The issue is that if the fashion industry is to reach the goals its leaders are vowing to work toward in the proclamations, it must lead in the creation of a cultural change. At this point, as Chapters 4, 5, and 6 show, governance actions have been enacted with little effect. It is because the culture of overconsumption persists. Government actions dare not impinge its commercial volumes, safeguarding its marginal growth.

Apart from symbolic government action, there are two additional social reasons as to why the culture of overconsumption persists despite ecological admonitions and calls to change purchasing habits. One is the economic logic of its business model reliant on the success of increasing sales above all. The other is that the consumption, and entire performance of showcasing new clothes, as people create their digital fantasy personas, acts as an antidote to the massive problem of social inequality. It helps those not as affluent, feel that they look as though they are. It provides a false feeling of democracy and equality (Krause, 2018).

In a culture that keeps on celebrating wealth, and wealth measures are constantly going up – the rich are getting mega rich, showcasing ostentatious lifestyles – all citizens feel the pressure to showcase maximum personal wealth. Sociologist Gunnar Trumbull (2018) argues that individuals are culturally "brainwashed" to feel that overconsumption and pretentious display of material worth is the way to manifest social status. It is these cultural forces, and the associated market incentives to participate in the culture of "display consumerism" (Carolan, 2005; Stearns, 2006, 2009) or "commodity fetishism" (Freedman, 2015; Harribey, 2005), that are behind the paradoxes analyzed in this book. Chief among them is the reality that while activism around the ecological footprint of fast fashion increases, so does fast fashion innovation, product development, and aggregate consumption.

It was 15 years ago that Barnes, Lea-Greenwood and Joergens (2006) noted that, as awareness of the environmental problems of apparel consumption started to grow, fashion buyers were unwilling to change their consumption habits. The research covered in this book indicates that still to be the case, helped by the massive and effective greenwashing response of fashion conglomerates. Yet, greenwashing can be handled and should be handled for it is nothing but false advertising. There are already laws against false advertising that have been executed the world over for the better part of the modern industrial age (Nehf, 2018; Petty, 1997). They are implemented on per-case bases, when consumer watch groups, individual citizens, or competing firms file action against a firm, be it a producer or its advertising arm, for misleading, false, and/or deceptive messaging.

The reason why there is no evidence of false advertising charges in fashion greenwashing is also cultural. Society likes sustainability messaging. For the reasons tracked in Chapter 3, the main fashion buying demographic – young people – like to hear, read, and believe that clothes manufacturers are making strides toward sustainability. At large, they do not understand what sustainability means, interpreting the concept on individual bases, internalizing the positive messaging with little regard for veracity. The problem is exacerbated by the fact that even the emerging governance structures are helping consumers to accept what is becoming a culture of greenwashing propaganda. By providing their own bombastic messaging of policy innovations, as explained in Chapter 6, and touting impressive figures as "goals," yet not offering any substantial details on how such policy innovation is to be implemented and goals be achieved, governments accomplish one thing – showcase willingness to acknowledge a problem. Yet, how they will legislate its solution is still to be seen.

The United Nation's Fashion Industry Charter on Climate Change is the main policy goal platform from which future legal frameworks for sustainability measurement and compliance will emerge (UNFCCC, 2018). These goals include a "call" for industrial reduction of green gas house emissions of 30% by 2030. As discussed earlier, such messaging is problematic because it: (a) focuses on only air pollution, not the main ecological issue in apparel, which is chemical effluent discharge and (b) speaks to the future. The mere concept of by "2030" offers a comforting notion to consumers that solutions are not essential to act on today. The underlying message is that their daily consumption habits need not change at present.

Perhaps because of such non-committal, yet showy proclamations, the main global apparel brands and most-noted fashion designers, have signed the chapter. They are all offering their own sustainability reports on their social media podiums, using the Charter as a platform from which to showcase a commitment to change. It is the most efficient of marketing tactics because it does provide positive information to consumers with reassurance that the industry is working toward change. However, the positive messages of change are ambiguous, lack details, and in some cases, are misleading if not outright false. For example, *H&M's* sustainability report, easily accessible on the firm's website, claims that 57% of materials used are recyclable. Yes, they very well could be, but most likely will not get recycled, much less be included in circular production, as Chapters 3 and 4 explain.

Another impressive figure included in the *H&M* sustainability report is the claim that goals are in place to have 100% "recycled" and "sustainably sourced" materials by 2030. As technology is developing in more efficient recycling, such a goal is important. Yet, as explained in previous chapters, technologies to reach that goal are not yet available to scale, much less at costs that would fit with the price points of *H&M's* product. When it comes to the most environmentally taxing input – cotton – the report promises to use 100% "sustainably sourced," "recycled" (not feasible at present, even with the innovations tracked in Chapter 3), "better cotton" (under the compliance of the Better Cotton Initiative) or "certified organic." As explained in detail in this book and in its prequel Anguelov (2015), none of these options, except for "better cotton," and only partially, can be called "sustainable."

The most disturbing claim in the *H&M* sustainability report is the promise to become "climate positive" by 2040, with the explanation that being "climate positive" means to "capture" more CO_2 emissions than its "core supply chain" emits. This is absolutely impossible to achieve, as even *H&M* would not have full understanding of who is in its "core supply chain" and would have no jurisdictional or oversight power to demand suppliers to change their energy use. Even if the firm tries to place such demands as conditions, it would be impossible to expect producers in its core sourcing markets – Bangladesh and Ethiopia for example, as they were discussed in some detail in this book – to comply because renewable source energy options there are not available. They are barely available in the developed world and would most likely remain slow to penetrate large scale utility energy production (Anguelov & Dooley, 2019). That is the case because there is still no efficient technology to store energy generated from renewable sources (Al-Ghussain et al., 2020).

To get back to fashion supply chains, the majority of energy is used in manufacturing factories and comes from coal (Connell & LeHew, 2020; Pattanayak, 2020). Then the transportation link, mainly reliant on coal and oil powered ocean liners, rail and trucking, continues to increase the carbon footprint. There is not only no technology currently being discussed for solar or wind-powered ocean liners, cargo trains, or long-distance cargo trucks, but if the experience with solar cars has taught us anything, it would suggest that even if such technology were in its development stages today, its adoption will be slow at best (Bennett & Vijaygopal, 2018; Patt, et al., 2019; Lane et al., 2018).

Promising a "carbon positive" future by 2040 is irresponsible at best, if not blatantly false advertising. Yet, *H&M* can defend such preposterous claims by simply stating that they are compliant with specific UN Development Goals, namely goals 6, 7, 12, 13, 14, 15, and 17. If readers wanted to understand what these goals are, they would have to go find for themselves on the UN website.

Other ambiguously worded goals in the Charter include the "aim" to work with signatories to establish a "decarbonization" path, which involves reducing emissions from transports and production. This is actually a feasible goal when signatories opt for electric transportation options, and establish a shorter supply chain network that offers preferential treatment to suppliers that are geographically nearby. The problem is that the implementation of such incentives will come with costs and those costs would have to be borne by the producers. Precedent, as tracked in Chapters 4 and 5, suggests that they would ask for government subsidies to offset such costs, not pass on the new costs to their consumers in the form of higher prices. Therefore, it can be expected that aggregate consumption will not be affected.

Another goal specified by the Chapter is the implementation of "methodologies" from the Science Based Targets initiative (SBTi). SBTi is a consultancy. As already discussed throughout the book, the use of consultancies, that offer voluntary compliance certificates, may actually be aiding in greenwashing promotional platforms. In the case of SBTi, a quick google search reveals the initiative consults manufacturers across industrial sectors by evaluating their "sustainability goals."[1] It is not clear what is the oversight process of reaching such goals.

It is very common now for firms from all sectors to pledge to make sustainability changes. When, how, and in what ways, is becoming a matter of agreement between a firm and the consulting voluntary compliance facilitator it hires. Because these facilitators have an incentive to increase their revenue by pleasing their clients, it is a market structure prone to "capture," as is the political economy term.

Finally, the UN Fashion Industry Charter for Climate Action calls for "opening up a dialogue" between companies and consumers to "improve circularity" and bring fashion and politics together in order to drive "new legislations." As the industry is globally fragmented through the tier system, the implication is to work toward international environmental legislation for fashion production, which is a very challenging goal, as explained by the more recent work in the vast literature on international environmental politics (DeSombre, 2018; Fiorino, 2018; Mitchell et al., 2020; Panke, 2020). Elizabeth DeSombre, the Camilla Chandler Frost Professor of Environmental Studies at Wellesley College asks as the title of her book: *What is Environmental Politics*? This seemingly simple and straightforward question, to which one would expect there to be a well-established definition, is in actuality a complicated conundrum (DeSombre, 2020). The book offers a detailed explanation of the reality of the conflict between the core tenant of national sovereignty – non-infringement on national law by foreign governments – and the need for enforcement mechanisms of international environmental law.

Up to now, all such law has been subject to voluntary agreement among states that vow to comply. This platform leads to a collective action problem because the group of signatory countries must agree on policies for detection and punishment of violations, and also, collectively do it. DeSombre (2020) offers numerous examples of how this never happens. Nations found in violation of environmental agreements simply leave the agreement rather than implement corrective policies toward compliance. In light of such realities, in effect the UN Fashion Industry Charter on Climate Action "asks" individual politicians in different countries to run for office on fashion sustainability campaigns and/or to include fashion sustainability goals into their political pledges.

It is a noble call, yet, after reading this book, one would be apt to agree that it is akin to calling for politicians to run on platforms of supporting world peace. Of course, any politician can vow to support world peace. But none can be held accountable for not making it happen.

Still, the Charter is not to be dismissed. It is an important first step toward creating new legislation. It shows that there is supranational support for such political action, and that is a development in the global governance of apparel trade that is unprecedented. All such legislation previously has been on market access, trade mark infringement, and protectionism from competition (Anguelov, 2015). However, it does not address the second main factor that impedes the implementation of a change in fashion buying culture – consumerism. That is the case because, as explained in Chapter 5, there is no alternative to equating economic growth with increasing consumption. Most likely, there will not be any change in policy or social activism to develop an economic prosperity model that includes valuation for conservation and saving of resources as the world grapples with COVID-19. The focus will be on re-stimulating consumption to increase taxation (from that consumption) to help governments the world over that are trillions in debt as a function of stimulus packages.

Climbing out of economic recessions involves as a first step, increasing consumption. The promotion of consuming more will be hard to reconcile with a call for a change in personal fashion buying habits. As the economic down-turn of the COVID epidemic continues, one can expect a propensity to observe a "lipstick effect" in fashion and across other industrial sectors that retail fast moving consumer goods (FMCGs) (Hill et al., 2012; Netchaeva & Rees, 2016). As explained in Chapter 1, that is the propensity for the sales of relatively inexpensive items to increase because when buyers are on a budget, they look for affordable ways to achieve gratification from new purchases.

In a fashion culture of global commodity fetishism, driven by social media influencing, with an official government encouragement to spend, buyers will face increasing temptation to buy cheap clothes. In light of this reality, the options of changing the cultural trend of buying gratuitous amounts of clothes on impulse are not easy to operationalize. Unfortunately, there is not enough of a social movement to change purchasing habits.

When such a question is raised, the industry has a ready response, rebutting the proposal to limit item sales with vows of sustainability. At the end of the BBC documentary special, *The Sustainability Challenge*, tracked in Chapter 3, *H&M's* "head of sustainability" Giorgina Waltier is asked in the closing scene, as the discussion sumps up the facts that not enough technology nor capability exists to make *H&M* truly sustainable: "... will you (meaning *H&M*) offer less product?" Georgina Waltier answers: "We will not offer less product. We will offer better product!" End scene and film (BBC World News, 2019).

Better? Better how? By how much? Better for whom? In what context? Those are the questions that need to be asked in order for transparency in production and circular initiatives to be established.

It is a hopeful time in the fashion industry, or at least it was until the onset of COVID-19, with the sector showing viable signs of willingness to transform. With those signs as guidance, I set off to write this book with a focus on the good news in Fashion. There is good news, and on-going innovations in new products and models, such as mass-customization, as well as governance, as tracked in Chapter 6. Yet, it is not enough.

The transformation toward sustainability, and even the definition of sustainability in fashion, is in its early stages. To develop, it needs market support, meaning a customer base willing to purchase its innovative products. It also needs government action. It is up to consumers to support the first and put pressure on the latter. Government is responsive to the concerns of its citizenry. As citizens, fashion buyers can exercise their political voice to call for policy change. The challenge is to create a global citizenry call, and for that, the on-going works of advocacy groups, non-profit organizations, and consumer watch groups are essential.

Note

1 for more, see: https://sciencebasedtargets.org

Appendix

List of All Nations

Country	Region	Net FDI		HO2 Textile Industry		HO2 Chemical Pollution		Carbon Dioxide Damage		People on > $2		GNI per Capita	
		1991	2008	1991	2008	1991	2008	1991	2008	1991	2008	1991	2008
Albania	Europe	20000000.00	843676732.32	59.80	60.19	..	..	19843732.40	39752586.32	6.5	7.85	2,020	8,360
Bangladesh	Asia and Oceania	1390444.32	973108114.49	77.11	77.11	3.22	3.22	79649773.55	345626848.51	92.54	81.33	570	1,600
Belarus	Europe	..	2149200000.00	..	..	..	..	535498611.21	648504968.19	13.6	13.6	4,810	12,840
Bulgaria	Europe	55900000.00	8472194672.89	20.68	28.04	10.52	10.52	298443872.00	418912932.87	4.17	4.17	4,650	13,250
Cambodia	South-East Asia	33000000.00	794691393.09	6.83	59.35	33.51	33.51	2308689.09	35903716.23	77.85	57.83	650	1,960
Cape Verde	Africa	1199579.88	213833531.95	..	..	..	..	458073.23	2442717.34	40.22	40.22	1,210	3,510
China	South-East Asia	3453000000.00	94320092013.51	..	..	..	..	12669884165.04	54876987806.96	78.58069	36.2785	890	6,250
Czech Republic	Europe	564357920.25	8966891344.97	15.21	7.40	7.08	10.89	749901731.96	994212914.49	2	2	10,520	24,690
Dominican Republic	Latin American and the Caribbean	145000000.00	2884700000.00	73.07	73.07	2.34	2.34	50644576.51	190686067.27	14.48	12.28	2,730	8,060
Egypt, Arab Rep.	Africa	191000000.00	7574400000.00	31.11	31.11	13.88	13.88	393319999.78	1426884995.27	27.64	18.46	2,370	5,710
Estonia	Europe	80399561.21	875931161.99	23.62	8.78	6.72	8.42	135911735.02	154205510.63	2.81	2	6,970	20,360
Fiji	South-East Asia	11927695.13	332673303.37	38.56	38.56	4.25	4.25	3371418.99	12133993.07	..	..	2,400	4,600
Haiti	Latin American and the Caribbean	11800000.00	29800000.00	0.00	..	0.00	..	4983836.76	13518008.90	72.15	72.15	1,010	1,140
Hong Kong	South-East Asia	..	3082975861.90	..	..	..	..	144384682.67	349203731.44	..	..	17,950	46,260
India	Asia and Oceania	73537638.39	22807027033.51	..	..	..	..	3687287964.01	13350715403.94	81.70923053	75.59853385	910	3,000
Indonesia	South-East Asia	1482000000.00	3418723398.71	31.61	31.61	12.77	12.77	906746800.94	2798459962.87	56.85	59.99	1,390	3,740

Korea, Rep.	Asia and Oceania	-308800000.00	-10594700000.00	24.99	9.34	9.62	12.05	1329218581.19	4003584060.94	..	..	8,960	27,080
Lao PDR	South-East Asia	6900000.00	..	..	..		..	1264282.12	12955610.01	84.82	76.85	710	2,090
Latvia	Europe	27291249.00	1092000000.00	19.93	12.61	5.61	5.59	72398599.40	63241867.62	2	2	7,080	17,930
Lesotho	Africa	273587899.23	218041081.45	90.14	90.75	0.79	1.20	0.00	0.00	70.88	62.25	1,080	1,920
Lithuania	Europe	30175186.84	1383367895.16	23.30	19.33	5.66	7.57	120352212.15	123310729.71	15.18	2	9,080	18,940
Macao	South-East Asia	..	3494246374.57	..	..	..	..	5460232.92	26956234.15	..	..	17,370	56,760
Macedonia	Europe	..	612032086.19	..	..	..	..	56322166.69	95289762.87	3.53	5.3	5,330	10,780
Madagascar	Africa	13681239.65	85444105.00	59.93	58.95	11.71	12.38	5368618.28	24286832.91	88.43	89.61	660	1,030
Maldives	South-East Asia	6500000.00	15427006.64	..	..	..	..	861177.68	6683740.47	..	..	1,970	5,370
Mauritius	Africa	6516610.70	325298218.48	..	..	..		7604015.65	28532671.19	..	..	4,730	12,780
Nepal	Asia and Oceania	19160171.09	995123.93	38.66	38.66	5.78	5.78	4800607.47	25111290.43	88.12	77.57	550	1,120
Northern Mariana Islands	South-East Asia	..	..	..	..	..	..	0.00	0.00	..	..	..	..
Pakistan	Asia and Oceania	262151741.78	5389000000.00	..	..	..	..	341008036.70	1113155321.71	88.18	60.31	1,260	2,570
Sri Lanka	South-East Asia	43825645.32	690500000.00	43.56	43.56	8.96	8.96	20301805.63	103821017.56	49.5	39.74	1,520	4,400
Tunisia	Africa	122212258.18	2600674976.50	..	..	..	..	77396053.25	194547342.48	20.39	12.82	2,920	7,530
Turkey	Asia and Oceania	783000000.00	15414000000.00	30.27	35.66	8.34	9.77	747941972.93	2061464340.61	9.84	8.23	2,920	7,530

References

Acharya, A., & Rahman, Z. (2016). Place branding research: A thematic review and future research agenda. *International Review on Public and Nonprofit Marketing*, 13(3), 289–317.

Acharyya, J. (2009). FDI, growth and the environment: evidence from India on CO2 emission during the last two decades. *Journal of Economic Development*, 34 (1), 43–58.

Adhikari, R. & Yamamoto, Y. (2007). Textile and clothing industry: Adjusting to the post-quota world. Chapter 1 in Kjöllerström, M. & O'Connor, D. (eds.) *Industrial Development for the 21st Century*. Chicago, IL: University of Chicago Press.

Akamatsu, K. (1962). A historical pattern of economic growth in developing countries. *The Developing Economies*, 1, 3–25.

Al-Ghussain, L., Samu, R., Taylan, O., & Fahrioglu, M. (2020). Sizing renewable energy systems with energy storage systems in microgrids for maximum cost-efficient utilization of renewable energy resources. *Sustainable Cities and Society*, 55, 102059–102076.

Al-Mulali, U., Weng-Wai, C., Sheau-Ting, L., & Mohammed, A. H. (2015). Investigating the environmental Kuznets curve (EKC) hypothesis by utilizing the ecological footprint as an indicator of environmental degradation. *Ecological Indicators*, 48, 315–323.

Alam, M. G. M., Allinson, G., Stagnitti, F., Tanaka, A., & Westbrooke, M. (2002). Arsenic contamination in Bangladesh groundwater: a major environmental and social disaster. *International Journal of Environmental Health Research*, 12(3), 235–253.

Alden, D. L., Steenkamp, J. B. E., & Batra, R. (1999). Brand positioning through advertising in Asia, North America, and Europe: The role of global consumer culture. *Journal of Marketing*, 63(1), 75–87.

Alden, C., Large, D., & Oliveira, R. S. D. (2008). *China Returns to Africa: A Rising Power and a Continent Embrace*. London, UK: Hurst Publishing.

Ali, I., Kim, S. R., Kim, S. P., & Kim, J. O. (2016). Recycling of textile wastewater with a membrane bioreactor and reverse osmosis plant for sustainable and cleaner production. *Desalination and Water Treatment*, 57(57), 27441–27449.

Allenby, G. M., Jen, L., & Leone, R. P. (1996). Economic trends and being trendy: The influence of consumer confidence on retail fashion sales. *Journal of Business & Economic Statistics*, 14(1), 103–111.

Amed, I., Berg A., Brantberg, L., Hedrich, S. (2017). *The state of fashion*. McKinsey & Company Report, December, available at: https://www.mckinsey.com/industries/retail/our-insights/the-state-of-fashion.

Anas, A. (July 17, 2015). *Textile plants are Dhaka's water problem – and also its solution. Citylab.com.* available at: https://www.citylab.com/life/2015/07/textile-plants-are-dhakas-water-problem-and-also-its-solution/398822/.

Andersen, M., & Skjoett-Larsen, T. (2009). Corporate social responsibility in global supply chains. *Supply Chain Management: An International Journal*, 14(2), 75–86.

Andriamananajara, S., Dean, J. & Springer, D. (2004). *Trading Apparel: Developing Countries in 2005.* Germany: Kiel Institute of World Economics.

Angelis-Dimakis, A., Alexandratou, A., & Balzarini, A. (2016). Value chain upgrading in a textile dyeing industry. *Journal of Cleaner Production*, 138, 237–247.

Anguelov, N. (2015). *The Dirty Side of the Garment Industry: Fast Fashion and Its Negative Impact on Environment and Society.* Boca Raton: CRC Press.

Anguelov, N. & Dooley, W.F. (2019). Renewable portfolio standards and policy stringency: An assessment of implementation and outcomes. *Review of Policy Research*, 36(2), 195–216.

Anholt, S. (2006). *Brand New Justice: How Branding of Places and Products Can Help the Developing World.* Elsevier: Butterworth-Heinemann.

Anholt, S. (2007). *Competitive Identity: The New Brand Management for Nations, Cities and Regions.* New York: Palgrave Macmillan.

Anner, M. (2009). Two logics of labor organizing in the global apparel industry. *International Studies Quarterly*, 53(3), 545–570.

Anner, M. (2011). The impact of international outsourcing on unionization and wages: Evidence from the apparel export sector in Central America. *Industrial and Labor Relations Review*, 64(2), 305–322.

Anner, M. (2019). Predatory purchasing practices in global apparel supply chains and the employment relations squeeze in the Indian garment export industry. *International Labour Review*, 158(4), 705–727.

Antonakakis, N., & Collins, A. (2018). A suicidal Kuznets curve?. *Economics Letters*, 166, 90–93.

Antweiler, W., Copeland, B. R., & Taylor, M. S. (2001). Is free trade good for the environment?. *American Economic Review*, 91(4), 877–908.

Apergis, N., & Ozturk, I. (2015). Testing environmental Kuznets curve hypothesis in Asian countries. *Ecological Indicators*, 52, 16–22.

Appelbaum, R. (2005). *TNCs and the removal of textiles and clothing quotas. United Nations Conference on Trade and Development Report*, Geneva: Switzerland, available at: https://unctad.org/en/Docs/iteiia20051_en.pdf.

Appelbaum, R., Bonacich, E., & Quan, K. (2005). The end of apparel quotas: A faster race to the bottom? Santa Barbara, CA: University of Santa Barbara, *Center for Global and International Studies*, available at: https://escholarship.org/uc/item/0q40t681.

Arnarson, Ö. & Hardarson, T. (2014). *Trend Beacons.* New York: Grasshopper Films.

Arora, A., Jaju, A., Kefalas, A. G., & Perenich, T. (2004). An exploratory analysis of global managerial mindsets: A case of US textile and apparel industry. *Journal of international Management*, 10(3), 393–411.

Arrow, K., Bolin, B., Costanza, R., Dasgupta, P., Folke, C., Holling, C. S., Jansson, B., Levin, S., Maler, K., Perrings, C., & Pimentel, D. (1996). Economic growth, carrying capacity, and the environment. *Environment and Development Economics*, 1(1), 104–110.

Arvidsson, A. (2006). *Brands: Meaning and Value in Media Culture*. London: Routledge.

Ash, T. G. (2016). *Free Speech: Ten Principles for a Connected World*. New Haven, CT: Yale University Press.

Au, K. F., & Wong, M. C. (2007). Textile and clothing exports of developed & developing countries: An analysis under the restrictive trade regime. *Journal of the Textile Institute*, 98(5), 471–478.

Auty, S., & Elliott, R. (1998). Fashion involvement, self-monitoring and the meaning of brands. *Journal of Product & Brand Management*, 7(2), 109–123.

Azevedo, A., & Farhangmehr, M. (2005). Clothing branding strategies: Influence of brand personality on advertising response. *Journal of Textile and Apparel, Technology and Management*, 4(3), 1–13.

Azmeh, S., & Nadvi, K. (2013). 'Greater Chinese' global production networks in the Middle East: The rise of the Jordanian garment industry. *Development and Change*, 44(6), 1317–1340.

Azuma, N., & Fernie, J. (2003). Fashion in the globalized world and the role of virtual networks in intrinsic fashion design. *Journal of Fashion Marketing and Management: An International Journal*, 7(4), 413–427.

Bafana, A., Devi, S. S., & Chakrabarti, T. (2011). Azo dyes: Past, present and the future. *Environmental Reviews*, 19(NA), 350–371.

Baffes, J. (2011). Cotton subsidies, the WTO, and the "cotton problem". *The World Economy*, 34(9), 1534–1556.

Bailey, F. & Barbato, R. (2012). *In Vogue: The Editor's Eye*. Los Angeles, CA: HBO Films.

Bair, J., & Gereffi, G. (2004). Upgrading, uneven development, and jobs in the North American apparel industry. In: Milberg, W. (eds) *Labor and the Globalization of Production* (pp. 58–87). London: Palgrave Macmillan.

Bansal, N., & Kanwar, S. S. (2013). Peroxidase (s) in environment protection. *The Scientific World Journal*, 2013. doi: 10.1155/2013/714639.

Banuri, T. (1998). Pakistan: Environmental impact of cotton production and trade. *Winnipeg, Canada. International Institute for Sustainable Development*. available at: https://www.iisd.org/sites/default/files/publications/pk_Banuri.pdf.

Barnes, L., & Lea-Greenwood, G. (2010). Fast fashion in the retail store environment. *International Journal of Retail & Distribution Management*, 38(10), 760–772.

Barnes, L., Lea-Greenwood, G., & Joergens, C. (2006). Ethical fashion: Myth or future trend?. *Journal of Fashion Marketing and Management: An International Journal*, 10(3), 360–371.

Barnett, J. M., Grolleau, G., & Harbi, S. E. (2010). The fashion lottery: Cooperative innovation in stochastic markets. *The Journal of Legal Studies*, 39(1), 159–200.

Bartsch, F., Diamantopoulos, A., Paparoidamis, N., & Chumpitaz, R. (2016). "Global brand ownership: The mediating roles of consumer attitudes and brand identification." *Journal of Business Research*, 69(9), 3629–3635.

Batra, R., Ramaswamy, V., Alden, D. L., Steenkamp, J. B. E., & Ramachander, S. (2000). Effects of brand local and nonlocal origin on consumer attitudes in developing countries. *Journal of Consumer Psychology*, 9(2), 83–95.

Beath, J., & Katsoulacos, Y. (1991). *The Economic Theory of Product Differentiation*. Cambridge, UK: Cambridge University Press.

BBC. (11 August, 2018) The Price of Fast Fashion, https://www.bbc.co.uk/programmes/n3ct5bcf

BBC. (26 September, 2019). *Click*. Segment: The new high-tech way to recycle clothes.

BBC World News. (27 April, 2019). Protecting Our Planet: The Sustainability Challenge.

BBC World News. (28 May, 2020). *Newsbeat*. Segment: Killer Kicks.

Bell, D., Gallino, S., Moreno, A., Yoder, J., & Ueda, D. (2018). The store is dead-long live the store. *MIT Sloan Management Review*, 59(3), 59–66.

Bell, J., McNaughton, R., & Young, S. (2001). 'Born-again global'firms: An extension to the 'born global'phenomenon. *Journal of International Management*, 7(3), 173–189.

Bellemare, M. F., Masaki, T., & Pepinsky, T. B. (2017). Lagged explanatory variables and the estimation of causal effect. *The Journal of Politics*, 79(3), 949–963.

Bendoni, W. K. (2017). *Social Media for Fashion Marketing: Storytelling in a Digital World*. London: Bloomsbury Publishing.

Bennett, R., & Vijaygopal, R. (2018). Consumer attitudes towards electric vehicles. *European Journal of Marketing*, 52(3/4), 499–527.

Bhaduri, G. & Ha-Brookshire, J. E. (2011). Do transparent business practices pay? Exploration of transparency and consumer purchase intention. *Clothing and Textiles Research Journal*, 29(2), 135–149.

Bhutiani, R., Varun, A. F. T., & Khushi, R. (2018). Evaluation of water quality of River Malin using water quality index (WQI) at Najibabad, Bijnor (UP) India. *Environment Conservation Journal*, 19(1&2), 191–201.

Bick, R., Halsey, E., & Ekenga, C. C. (2018). The global environmental injustice of fast fashion. *Environmental Health*, 17(1), article number 92. doi: 10.1186/s12940-018-0433-7

Bikhchandani, S., Hirshleifer, D., & Welch, I. (1992). A theory of fads, fashion, custom, and cultural change as informational cascades. *Journal of political Economy*, 100(5), 992–1026.

Bilyaeu, N. (15 January 2018). The true story of Gianni Versace's murder. *Town & Country Magazine*, available at: https://www.townandcountrymag.com/style/fashion-trends/a15045343/gianni-versace-murder-death-true-story/

Bini, L., & Bellucci, M. (2020). Business model disclosure in sustainability reporting: Two case studies. In Bini, L. & Bellucci, M. (eds)*Integrated Sustainability Reporting* (pp. 117–150). Switzerland, Cham: Springer.

Binnemans, K., Jones, P. T., Blanpain, B., Van Gerven, T., Yang, Y., Walton, A., & Buchert, M. (2013). Recycling of rare earths: A critical review. *Journal of Cleaner Production*, 51, 1–22.

Birkey, R. N., Guidry, R. P., Islam, M. A., & Patten, D. M. (2018). Mandated social disclosure: An analysis of the response to the California Transparency in Supply Chains Act of 2010. *Journal of Business Ethics*, 152(3), 827–841.

Birnbaum, D. (2005). *Birnbaum's Global Guide to Winning the Great Garment War*. 5th Edition. Hong Kong: Third Horizon Press.

Birnbaum, D. (2008). *Crisis in the 21st Century Garment Industry and Breakthrough Unified Strategy*. New York: Fashiodex.

Birnbaum, D. (2015). *Birnbaum's Global Guide to Agents and Buying Offices*. New York: Fashiondex.

Birtwistle, G., & Moore, C. M. (2007). Fashion clothing-where does it all end up?. *International Journal of Retail and Distribution Management*, 35(3), 210–216.

Bismar, A. (10 May 2014). Review of Copenhagen Fashion Summit. greenstrategy.se, available at: http://www.greenstrategy.se/review-copenhagen-fashion-summit-2014-2/

Black, S. and Eckert, C.M. (2010). Developing considerate design: meeting individual fashion and clothing needs within a framework of sustainability. in Piller, F.T. and Tseng, M.M. (Eds), *Handbook of Research in Mass Customization and Personalization*, Vol. 2, World Scientific Publishing Company, Singapore, pp. 813–832.

Blaszczyk, R. L. (Ed.). (2011). *Producing fashion: Commerce, Culture, and Consumers*. Philadelphia: University of Pennsylvania Press.

Bly, S., Gwozdz, W., & Reisch, L. A. (2015). Exit from the high street: An exploratory study of sustainable fashion consumption pioneers. *International Journal of Consumer Studies*, 39(2), 125–135.

Boerman, S. C. (2020). The effects of the standardized Instagram disclosure for micro-and meso-influencers. *Computers in Human Behavior*, 103, 199–207.

Boockmann, B., & Dreher, A. (2003). The contribution of the IMF and the World Bank to economic freedom. *European Journal of Political Economy*, 19(3), 633–649.

Brautigam, D. (2008). Chinese business and African development: Flying geese or—hidden dragon?. In Alden, C., Large, D., & Soares de Oliveira, R. (eds) *China Returns to Africa: A Rising Power and a Continent Embrace*. London: Hurst Publications.

Bray, J., Johns, N., & Kilburn, D. (2011). An exploratory study into the factors impeding ethical consumption. *Journal of Business Ethics*, 98(4), 597–608.

Braconier, H., Norbäck, P. J., & Urban, D. (2005). Multinational enterprises and wage costs: Vertical FDI revisited. *Journal of International Economics*, 67(2), 446–470.

Bridson, K., & Evans, J. (2004). The secret to a fashion advantage is brand orientation. *International Journal of Retail & Distribution Management*, 32(8), 403–411.

Brismar, A. (May 10, 2014). Review of Copenhagen fashion summit. Greenstrategy.se, available at: https://www.greenstrategy.se/review-copenhagen-fashion-summit-2014-2/

Brockington, D. (2014). The production and construction of celebrity advocacy in international development. *Third World Quarterly* 35(1), 88–108.

Broderick, A., & Pickton, D. (2005). *Integrated Marketing Communications* 2nd edition. Harlow, UK: Pearson Education.
Browning, M., & Crossley, T. F. (2000). Luxuries are easier to postpone: A proof. *Journal of political Economy*, 108(5), 1022–1026.
Bruce, M. & Daly, L. (2004). Lean or agile? A solution for supply chain management in the textile and clothing industry. *International Journal of Operations & Production Management*, 24(2), 151–170.
Bruce, M. & Daly, L. (2006). Buyer behavior for fast fashion. *Journal of Fashion Marketing and Management*, 10(3), 329–344.
Brunel, C. (2017). Pollution offshoring and emission reductions in EU and US manufacturing. *Environmental and Resource Economics*, 68(3), 621–641.
Brunel, C., & Levinson, A. (2016). Measuring the stringency of environmental regulations. *Review of Environmental Economics and Policy*, 10(1), 47–67.
Brüschweiler, B. J., Küng, S., Bürgi, D., Muralt, L., & Nyfeler, E. (2014). Identification of non-regulated aromatic amines of toxicological concern which can be cleaved from azo dyes used in clothing textiles. *Regulatory Toxicology and Pharmacology*, 69(2), 263–272.
Buckley, C. (1998). On the margins: Theorizing the history and significance of making and designing clothes at home. *Journal of Design History*, 11(2), 157–171.
Buscio, V., Crespi, M., & Gutiérrez-Bouzán, C. (2015). Sustainable dyeing of denim using indigo dye recovered with polyvinylidene difluoride ultrafiltration membranes. *Journal of Cleaner Production*, 91, 201–207.
Busse, M. (2010). On the growth performance of sub-Saharan African countries. *The Estey Centre Journal of International Law and trade Policy*, 11(2), 384–402.
Cachon, G. P., & Swinney, R. (2011). The value of fast fashion: Quick response, enhanced design, and strategic consumer behavior. *Management Science*, 57(4), 778–795.
Cao, H., Frey, L. V., Farr, C. A., & Gam, H. (2006). An environmental sustainability course for design and merchandising students. *Journal of Family and Consumer Sciences*, 98(2), 75–80.
Caric, K. (25 September, 2019). 10+ ethical fashion certifications you need to know. *ELUXE Magazine*, available at: https://eluxemagazine.com/magazine/ethical-fashion-certifications/.
Carley, S., Nicholson-Crotty, S., & Miller, C. J. (2017). Adoption, reinvention and amendment of renewable portfolio standards in the American states. *Journal of Public Policy*, 37(4), 431–458.
Carley, S., Davies, L. L., Spence, D. B., & Zirogiannis, N. (2018). Empirical evaluation of the stringency and design of renewable portfolio standards. *Nature Energy* 3, 754–763.
Carmody, P. R., & Owusu, F. Y. (2007). Competing hegemons? Chinese versus American geo-economic strategies in Africa. *Political Geography*, 26(5), 504–524.
Caro, F., & Gallien, J. (2012). Clearance pricing optimization for a fast-fashion retailer. *Operations Research*, 60(6), 1404–1422.

Caro, F., & Martínez-de-Albéniz, V. (2015). Fast fashion: Business model overview and research opportunities. In Agrawal, N. & Smith, S. (eds) *Retail Supply Chain Management*, 2nd. Edition (pp. 237–264). Springer, Boston, MA.
Carolan, M. (2005). The conspicuous body: Capitalism, consumerism, class and consumption. *Worldviews: Global Religions, Culture, and Ecology*, 9(1), 82–111.
Carpenter, J. M., & Fairhurst, A. (2005). Consumer shopping value, satisfaction, and loyalty for retail apparel brands. *Journal of Fashion Marketing and Management: An International Journal*, 9(3), 256–269.
Cayla, J., & Arnould, E. J. (2008). A cultural approach to branding in the global marketplace. *Journal of International Marketing*, 16(4), 86–112.
Cervellon, M. C., Carey, L., & Harms, T. (2012). Something old, something used: Determinants of women's purchase of vintage fashion vs second-hand fashion. *International Journal of Retail & Distribution Management*, 40(12), 956–974.
Chapman, P. M., Bailey, H., & Canaria, E. (2000). Toxicity of total dissolved solids associated with two mine effluents to chironomid larvae and early life stages of rainbow trout. *Environmental Toxicology and Chemistry: An International Journal*, 19(1), 210–214.
Chase-Dunn, C. (1975). The effects of international economic dependence on development and inequality: A cross-national study. *American Sociological Review*, 40(6), 720–738.
Chaturvedi, S., & G. Nagpal. (2003). WTO and product related environmental standards: Emerging issues and policy options. *Economic and Political Weekly*, 38(1), 66–74.
Chaudhuri, S., & Mukhopadhyay, U. (2014). *Foreign Direct Investment in Developing Countries*. Dordrecht, Germany: Springer.
Chavalitsakulchai, P., Kawakami, T., Kongmuang, U., Vivatjestsadawut, P., & Leongsrisook, W. (1989). Noise exposure and permanent hearing loss of textile workers in Thailand. *Industrial Health*, 27(4), 165–173.
Cheang, S., & Kramer, E. (2017). Fashion and East Asia: Cultural translations and East Asian perspectives. *International Journal of Fashion Studies*, 4(2), 145–155.
Chen, C. (6 June, 2018). *Fashion got woke. But at what cost?*. businessoffashion.com, available at: https://www.businessoffashion.com/articles/professional/fashion-got-woke-but-at-what-cost.
Chen, E. K. (1997). The total factor productivity debate: determinants of economic growth in East Asia. *Asian-Pacific Economic Literature*, 11(1), 18–38.
Chen, H. L., & Burns, L. D. (2006). Environmental analysis of textile products. *Clothing and Textiles Research Journal*, 24(3), 248–261.
Chen, Q., & Zhao, T. (2016). The thermal decomposition and heat release properties of the nylon/cotton, polyester/cotton and Nomex/cotton blend fabrics. *Textile Research Journal*, 86(17), 1859–1868.
Cherniwchan, J., Copeland, B. R., & Taylor, M. S. (2017). Trade and the environment: New methods, measurements, and results. *Annual Review of Economics*, 9, 59–85.
Chi, T. (2015). Consumer perceived value of environmentally friendly apparel: An empirical study of Chinese consumers. *The Journal of the Textile Institute*, 106(10), 1038–1050.

Cho, E., Gupta, S., & Kim, Y. K. (2015). Style consumption: Its drivers and role in sustainable apparel consumption. *International Journal of Consumer Studies*, 39(6), 661–669.

Choi, T. M. (2013). Optimal return service charging policy for a fashion mass customization program. *Service Science*, 5(1), 56–68.

Choi, T. M., Liu, S. C., Tang, C. S., & Yu, Y. (2011). A cross-cluster and cross-region analysis of fashion brand extensions. *Journal of the Textile Institute*, 102(10), 890–904.

Chung, W., & Alcácer, J. (2002). Knowledge seeking and location choice of foreign direct investment in the United States. *Management Science*, 48(12), 1534–1554.

Claudio, L. (2007). Waste couture: Environmental impact of the clothing industry. *Environmental Health Perspectives*, 115(9), A449–A454.

Cline, E. (2012). *Overdressed: The Shockingly High Cost of Cheap Fashion*. Penguin Group, New York, NY.

Cline, E. (April 5, 2018). *H&M's woes mean fast fashion is getting worse, not better.* latimes.com, avaialbe at: http://www.latimes.com/opinion/op-ed/la-oe-cline-hm-fast-fashion-20180405-story.html.

Clodfelter, R. (2015). *Retail Buying: From Basics to Fashion*. London: Bloomsbury Publishing.

Cohen, A. K. (2012). Designer collaborations as a solution to the fast-fashion copyright dilemma. *Chicago-Kent Journal of Intellectual Property*, 11(2), 172–185.

Cohen, M. J. (2014). When sustainability bites back: cautionary lessons from the field of public health. *Sustainability: Science, Practice, and Policy*, 10(2), 1–3

Cole, M. A., & Elliott, R. J. (2003). Determining the trade-environment composition effect: the role of capital, labor and environmental regulations. *Journal of Environmental Economics and Management*, 46(3), 363–383.

Collins, M. B., Munoz, I., & JaJa, J. (2016). Linking 'toxic outliers' to environmental justice communities. *Environmental Research Letters*, 11(1), 1–9.

Colucci, M., Montaguti, E., & Lago, U. (2008). Managing brand extension via licensing: An investigation into the high–end fashion industry. *International Journal of Research in Marketing*, 25(2), 129–137.

Connolly, C. B. (2015). Sustainability in the Apparel Industry: Improving How Companies Assess and Address Environmental Impacts Through a Revised Higg Index Facility Module, Master's thesis, Harvard Extension School.

Connell, K. Y. H., & LeHew, M. L. (2020). Fashion: An unrecognized contributor to climate change. In Marcketti, S. & Karpova, E. (eds) *The Dangers of Fashion: Towards Ethical and Sustainable Solutions*, pp. 71–86. London, UK: Bloomsbury Publishing.

Cook, D. T., & Kaiser, S. B. (2004). Betwixt and be tween: Age ambiguity and the sexualization of the female consuming subject. *Journal of Consumer Culture*, 4(2), 203–227.

Copeland, B. R., & Taylor, M. S. (2004). Trade, growth, and the environment. *Journal of Economic Literature*, 42(1), 7–71.

Corvellec, H. (2016). A performative definition of waste prevention. *Waste Management*, 52, 3–13.

Craik, J. (2019). "Feeling premium": Athleisure and the material transformation of sportswear. In Jenss, H. & Hofmann, V. (eds). *Fashion and Materiality: Cultural Practices in Global Contexts*, London: Bloomsbury.
Crane, D. (1999). Diffusion models and fashion: A reassessment. *The Annals of the American Academy of Political and Social Science*, 566(1), 13–24.
Crewe, L. (2017) *The Geographies of Fashion: Consumption, Space and Value.* London: Bloomsbury.
Cui, Y., & Lu, C. (2018). Are China's unit labour costs still competitive? A comparison with ASEAN countries. *Asian-Pacific Economic Literature*, 32(1), 59–76.
Daley, P. (2013). Rescuing African bodies: Celebrities, consumerism and neoliberal humanitarianism. *Review of African Political Economy*, 40(137), 375–393.
Danigelis, A. (2018, February 28). Report reveals global apparel and footwear industries' emissions. environmentalleader.com, available at: https://www.environmentalleader.com/2018/02/apparel-and-footwear-emissions/
Dascalu, T., Acosta-Ortiz, S. E., Ortiz-Morales, M., & Compean, I. (2000). Removal of the indigo color by laser beam-denim interaction. *Optics and Lasers in Engineering*, 34(3), 179–189.
Dasgupta, J., Sikder, J., Chakraborty, S., Curcio, S., & Drioli, E. (2015). Remediation of textile effluents by membrane-based treatment techniques: A state of the art review. *Journal of Environmental Management*, 147, 55–72.
Dastrup, S. R., Zivin, J. G., Costa, D. L., & Kahn, M. E. (2012). Understanding the Solar Home price premium: Electricity generation and "Green" social status. *European Economic Review*, 56(5), 961–973.
Daugherty, T., Eastin, M. S., & Bright, L. (2008). Exploring consumer motivations for creating user-generated content. *Journal of interactive advertising*, 8(2), 16–25.
Davis, D. (2000). Understanding international trade patterns: advances of the 1990s. *Integration & Trade*, 4, 61–79.
de Abreu, M. C. S., de Castro, F., de Assis Soares, F., & da Silva Filho, J. C. L. (2012). A comparative understanding of corporate social responsibility of textile firms in Brazil and China. *Journal of Cleaner Production*, 20(1), 119–126.
De Bruyns, S. (1997). Explaining the environmental Kuznets curve: Structural change and international agreements in reducing sulphur emissions. *Environment and Development Economics*, 2(4), 485–503.
DeHart, J. L., & Soulé, P. T. (2000). Does I= PAT work in local places?. *The Professional Geographer*, 52(1), 1–10.
Dem S., Cobb, J.M., & Mullins, D.E. (2007). Pesticides residues in soil and water from four cotton growing areas of Mali, West Africa. *Journal of Agricultural, Food, and Environmental Sciences*, 1(1), 1–12.
De Mooij, M., & Hofstede, G. (2010). The Hofstede model: Applications to global branding and advertising strategy and research. *International Journal of Advertising*, 29(1), 85–110.
DeSombre, E. R. (2018). *Why Good People Do Bad Environmental Things*. Oxford, UK: Oxford University Press.
DeSombre, E. R. (2020). *What is Environmental Politics?*. New York: John Wiley & Sons.

Desore, A., & Narula, S. A. (2018). An overview on corporate response towards sustainability issues in textile industry. *Environment, Development and Sustainability*, 20(4), 1439–1459.

Dewsnap, B. and Hart, C. (2004). Category management: A new approach for fashion marketing?. *European Journal of Marketing*, 38(7), 809–834.

Dey, S., & Islam, A. (2015). A review on textile wastewater characterization in Bangladesh. *Resources and Environment*, 5(1), 15–44.

Dickson, M. A., Lennon, S. J., Montalto, C. P., Shen, D., & Zhang, L. (2004). Chinese consumer market segments for foreign apparel products. *Journal of Consumer Marketing*, 21(5), 301–317.

Dietz, T., Stern, P. C., & Guagnano, G. A. (1998). Social structural and social psychological bases of environmental concern. *Environment and Behavior*, 30(4), 450–471.

Dietz, T., Ostrom, E., & Stern, P. C. (2003). The struggle to govern the commons. *Science*, 302(5652), 1907–1912.

Dinda, S. (2004). Environmental Kuznets curve hypothesis: A survey, *Ecological Economics*, 49(4), 431–455.

Directive, E. C. (2008). Directive 2008/98/EC of the European parliament and of the council of 19 November 2008 on waste and repealing certain directives. *Official Journal of the European Union L*, 312(3).

Douglas, S. P., Samuel, C., & Nijssen, E. J. (2001). "Integrating branding strategy across markets: Building international brand architecture." *Journal of International Marketing*, 9(2), 97–114.

Dowd, B. M. (2008). Organic cotton in Sub-Saharan Africa. In Moseley, W. & Gray, L. (eds.) *Hanging by a Thread: Cotton, Globalization, and Poverty in Africa*. Athens, OH: Ohio University Press.

Dowlinga, M., & Cheang, C. T. (2000). Shifting comparative advantage in Asia: New tests of the "flying geese" model. *Journal of Asian Economics*, 11(4), 443–463.

Dowsett, S & Fares, M. (18 September, 2019). Garments for lease: 'rental' apparel brings new wrinkles for retail stores. *Reuters Business News*, available at: https://www.reuters.com/article/us-retail-renting-focus/garments-for-lease-rental-apparel-brings-new-wrinkles-for-retail-stores-idUSKBN1W31CA.

Draper, D. R. (2015). Versailles' 73: American Runway revolution. *Journal of Contemporary African Art*, 2015(37), 94–103.

Drezner, D. W. (2001). Globalization and policy convergence. *International Studies Review*, 3(1), 53–78.

Duncan, T. R., & Everett, S. E. (1993). Client perceptions of integrated marketing communications. *Journal of Advertising Research*, 33(3), 30–40.

DuVerney, A. (2016). *The Battle of Versailles*. New York: HBO Films.

Easey, M. (Ed.). (2009). *Fashion Marketing*. New York: John Wiley & Sons.

Easterly, W. (2005). What did structural adjustment adjust? The association of policies and growth with repeated IMF and world bank adjustment loans. *Journal of Development Economics*, 76(1), 1–22.

Egels-Zandén, N., Hulthén, K., & Wulff, G. (2015). Trade-offs in supply chain transparency: The case of Nudie Jeans Co. *Journal of Cleaner Production*, 107, 95–104.

Ehrlich, P. R., & Holdren, J. P. (1971). Impact of population growth. *Science*, 171(3977), 1212–1217.
Ehrlich, P. R., & Ehrlich, A. H. (2010). The culture gap and its needed closures. *International Journal of Environmental Studies*, 67(4), 481–492.
Ekström, K. M., & Salomonson, N. (2014). Reuse and recycling of clothing and textiles: A network approach. *Journal of Macromarketing*, 34(3), 383–399.
Ellen MacCarther Foundation (2017). *A new textile economy: Redesigning fashion's future*, available at: https://www.ellenmacarthurfoundation.org/assets/downloads/A-New-Textiles-Economy_Full-Report_Updated_1-12-17.pdf
Elliott, R. J., & Shimamoto, K. (2008). Are ASEAN countries havens for Japanese pollution-intensive industry?. *World Economy*, 31(2), 236–254.
Elliott, R. & Wattanasuwan, K. (1998). "Brands as symbolic resources for the construction of identity." *International Journal of Advertising*, 17(2), 131–144.
English, B. (2013). *A Cultural History of Fashion in the 20th And 21st Centuries: From Catwalk to Sidewalk*. London: Bloomsbury.
EPA. (2016). *Advancing Sustainable Materials Fact Sheet: 2014*. Washington, DC. available at: https://www.epa.gov/sites/production/files/2016-11/documents/2014_smmfactsheet_508.pdf.
ESCAP (eds.) (2008). *Unveiling Protectionism: Regional Responses to Remaining Barriers in the Textiles and Clothing Trade*. Studies in Trade and Investment series. Bangkok, Thailand: United Nations Economic and Social Commission for Asia and the Pacific (ESCAP), number tipub2500.
European Parliament. (2017). *Resolution of 27 April 2017 on the EU Flagship Initiative on the Garment Sector. 2016/2140(INI)*. Brussels: European Parliament.
Fallers, L. A. (1954). A note on the "Trickle effect". *The Public Opinion Quarterly*, 18(3), 314–321.
Felice, K. B. (2011). Fashioning a solution for design piracy: Considering intellectual property law in the global context of fast fashion. *Syracuse Journal of International Law & Commmerce* 39(1), 219–247.
Ferrigno, S., Ratter, S. G., Ton, P., Vodouhê, D. S., Williamson, S., & Wilson, J. (2005). *Organic Cotton: A New Development Path for African Smallholders?* London, UK: International Institute for Environment and Development. available at: http://pubs.iied.org/pdfs/14512IIED.pdf.
FilmsMedia Group. (2016). *Zara: The Story of the World's Richest Man*. New York. ISBN: 978-1-64867-497-6.
Fiore, A. M., Lee, S. E., & Kunz, G. (2004). Individual differences, motivations, and willingness to use a mass customization option for fashion products. *European Journal of Marketing*, 38(7), 835–849.
Fiorino, D. J. (2018). *Can Democracy Handle Climate Change?*. New York: John Wiley & Sons.
Fischer, G. (2011). Understanding, fostering, and supporting cultures of participation. *Interactions*, 18(3), 42–53.
Fletcher, K. (2010). Slow fashion: An invitation for systems change. *Fashion Practice*, 2(2), 259–265.

Forney, J.C., Joo Park, E. and Brandon, L. (2005). Effects of evaluative criteria on fashion brand extension. *Journal of Fashion Marketing and Management*, 9(2), 156–165.

Franzen, A., & Meyer, R. (2009). Environmental attitudes in cross–national perspective: A multilevel analysis of the ISSP 1993 and 2000. *European Sociological Review*, 26(2), 219–234.

Freedman, D. (2015). Media policy fetishism. *Critical Studies in Media Communication*, 32(2), 96–111.

Frey, R. S., Gellert, P. K., & Dahms, H. F. (Eds.). (2018). *Ecologically Unequal Exchange: Environmental Injustice in Comparative and Historical Perspective*. Berlin: Springer.

Friedman, M. (2018). *Theory of the Consumption Function*. Princeton, NJ: Princeton University Press.

Friedman, T. L. (2006). *The World Is Flat: A Brief History of The Twenty-First Century*. New York: Palgrave Macmillan.

Fu, J., Nyanhongo, G. S., Gübitz, G. M., Cavaco-Paulo, A., & Kim, S. (2012). Enzymatic colouration with laccase and peroxidases: Recent progress. *Biocatalysis and Biotransformation*, 30(1), 125–140.

Ganguli, R., & Cook, D. R. (2018). Rare earths: A review of the landscape. *MRS Energy & Sustainability*, 5, doi: 10.1557/mre.2018.7.

Gans, H. (2008). *Popular Culture and High Culture: An Analysis and Evaluation of Taste*. New York: Basic books.

Garcia-Torres, S., Rey-Garcia, M., & Albareda-Vivo, L. (2017). Effective disclosure in the fast-fashion industry: from sustainability reporting to action. *Sustainability*, 9(12), 2256–2273.

Gardetti, M. Á. (2015). Making the connection between the United Nations global compact code of conduct for the textile and fashion sector and the sustainable apparel coalition higg index (2.0). In Muthu (ed) *Roadmap to Sustainable Textiles and Clothing* (pp. 59–86). Singapore: Springer.

Gardetti, M. A., & Muthu, S. S. (Eds.). (2020). *The UN Sustainable Development Goals for the Textile and Fashion Industry*. Singapore: Springer Verlag.

Garlapati, V. K. (2016). E-waste in India and developed countries: Management, recycling, business and biotechnological initiatives. *Renewable and Sustainable Energy Reviews*, 54, 874–881.

Gereffi, G. (1999). International trade and industrial upgrading in the apparel commodity chain. *Journal of International Economics*, 48(1), 37–70.

Geyer, R., Jambeck, J. R., & Law, K. L. (2017). Production, use, and fate of all plastics ever made. *Science Advances*, 3(7), e1700782–e1700787.

Ghaly, A. E., Ananthashankar, R., Alhattab, M. V. V. R., & Ramakrishnan, V. V. (2014). Production, characterization and treatment of textile effluents: A critical review. *Journal of Chemical Engineering & Process Technology*, 5(1), 1–19.

Gharfalkar, M., Court, R., Campbell, C., Ali, Z., & Hillier, G. (2015). Analysis of waste hierarchy in the European waste directive 2008/98/EC. *Waste Management*, 39, 305–313.

Ghemawat, P., Nueno, J. L., & Dailey, M. (2003). *ZARA: Fast Fashion*. Boston, MA: Harvard Business School.

Gill, A. R., Viswanathan, K. K., & Hassan, S. (2017). Is environmental Kuznets curve (EKC) still relevant?. *International Journal of Energy Economics and Policy*, 7(1), 156–165.

Gilmore, N. (16 January, 2018). Ready–to–waste: America's clothing crisis. *The Saturday Evening Post*, available at: https://www.saturdayeveningpost.com/2018/01/ready-waste-americas-clothing-crisis/.

Giovannini, S., Xu, Y., & Thomas, J. (2015). Luxury fashion consumption and Generation Y consumers. *Journal of Fashion Marketing and Management*, 19(1), 22–44.

Givhan, R. (2015). *The Battle of Versailles: The Night American Fashion Stumbled into the Spotlight and Made History*. New York: Flatiron Books.

Glover, S. (9 January, 2020). *Action taken against 'unsafe' garment factories*. Ecotextile.com, available at: https://www.ecotextile.com/2020010925525/social-compliance-csr-news/action-taken-against-unsafe-garment-factories.html.

Godart, F. C., & Mears, A. (2009). How do cultural producers make creative decisions? Lessons from the catwalk. *Social Forces*, 88(2), 671–692.

Goldsmith, R. E., Heitmeyer, J. R., & Freiden, J. B. (1991). Social values and fashion leadership. *Clothing and Textiles Research Journal*, 10(1), 37–45.

Goodfellow, W. L., Ausley, L. W., Burton, D. T., Denton, D. L., Dorn, P. B., Grothe, D. R., ... & Rodgers Jr, J. H. (2000). Major ion toxicity in effluents: A review with permitting recommendations. *Environmental Toxicology and Chemistry: An International Journal*, 19(1), 175–182.

Goodrum, A. (2015). The style stakes: Fashion, sportswear and horse racing in inter: War America. *Sport in History*, 35(1), 46–80.

Gordon, S., & Hsieh, Y. L. (Eds.). (2006). *Cotton: Science and Technology*. Sawston, UK: Woodhead Publishing.

Graham-Rowe, D. (2011). Robot tailoring: Stitched by the sewbot. *New Scientist*, 210(2817), 46–49.

Grainge, P. (2007). *Brand Hollywood: Selling Entertainment in a Global Media Age*. London: Routledge.

Gray, R. (2006). Social, environmental and sustainability reporting and organisational value creation? Whose value? Whose creation? *Accounting, Auditing & Accountability Journal*, 19(6), 793–819

Gray, J. V., Skowronski, K., Esenduran, G., & Johnny Rungtusanatham, M. (2013). The reshoring phenomenon: What supply chain academics ought to know and should do. *Journal of Supply Chain Management*, 49(2), 27–33.

Greene, W. H. (2008). *Econometric Analysis* 6th ed., Upper Saddle River, New Jersey, USA: Pearson/Prentice Hall.

Greenpeace International. (2012, November 20). Toxic Threads: The Big Fashion Stitch-up. available at: https://www.greenpeace.org/international/publication/6889/toxic–threads–the–big–fashion–stitch–up/.

Greenpeace International. (2017, September 18). Fashion at the Crossroads. available at: https://www.greenpeace.org/international/publication/6969/fashion–at–the–crossroads/.

Greenpeace International. (2018, July 12). Greenpeace Report: Clothing Industry Shows Progress in Cutting Hazardous Chemicals. available at: https://www.

greenpeace.org/international/press-release/17739/greenpeace-report-clothing-industry-shows-progress-in-cutting-hazardous-chemicals/.

Greer, L., Keane, S. E., & Lin, Z. (2010). *NRDC's Ten Best Practices for Textile Mills to Save Money and Reduce Pollution*. New York, NY: National Resource Defense Council. available at: http://www.nrdc.org/international/cleanbydesign/files/rsifullguide.pdf.

Grether, J. M., Mathys, N. A., & de Melo, J. (2009). Scale, technique and composition effects in manufacturing SO_2 emissions. *Environmental and Resource Economics*, 43(2), 257–274.

Grimes, P., & Kentor, J. (2003). Exporting the greenhouse: Foreign capital penetration and CO_2 emissions 1980 1996. *Journal of World-Systems Research*, 9(2), 261–275.

Gronwall, J., & Jonsson, A. C. (2017). Regulating effluents from India's textile sector: New commands and compliance monitoring for zero liquid discharge. *Law Environment & Development Journal*, 13(1), 15–29.

Guizzo, E. (2018). Your next t-shirt will be made by a robot. *IEEE Spectrum*, 55(1), 50–57.

Guo, X. (2013). "Living in a global world: Influence of consumer global orientation on attitudes toward global brands from developed versus emerging countries." *Journal of International Marketing*, 21(1), 1–22.

Gupta, A. & Gupta, R. (2019) Treatment and recycling of wastewater from pulp and paper mill. In Singh R., Singh R. (eds) *Advances in Biological Treatment of Industrial Waste Water and their Recycling for a Sustainable Future. Applied Environmental Science and Engineering for a Sustainable Future*. Singapore: Springer.

Halimi, M. T., Hassen, M. B., & Sakli, F. (2008). Cotton waste recycling: Quantitative and qualitative assessment. *Resources, Conservation and Recycling*, 52(5), 785–791.

Hall, C. (2000). *Boss Women: Anna Wintour*. London, UK: BBC Films.

Halvorsen, K. (2019). A retrospective commentary: How fashion blogs function as a marketing tool to influence consumer behavior: Evidence from Norway. *Journal of Global Fashion Marketing*, 10(4), 398–403.

Hampson, D. P., & McGoldrick, P. J. (2013). A typology of adaptive shopping patterns in recession. *Journal of Business Research*, 66(7), 831–838.

Han, Y. K., Morgan, G. A., Kotsiopulos, A., & Kang-Park, J. (1991). Impulse buying behavior of apparel purchasers. *Clothing and Textiles Research Journal*, 9(3), 15–21.

Hanbury, M. (26 May, 2019). Millennials' attitudes toward clothing ownership are brigning about a major change in the fashion industry. *Business Insider*, available at: https://www.businessinsider.com/millennials-renting-more-clothes-threatens-hm-zara-forever-21-2019-5

Harribey, J. M. (2005). Wealth and value: An incompatible couple. *L'Homme et la Société*, (2), 27–46.

Harris, F., Roby, H., & Dibb, S. (2016). Sustainable clothing: Challenges, barriers and interventions for encouraging more sustainable consumer behaviour. *International Journal of Consumer Studies*, 40(3), 309–318.

Hartmann, W. R. (2006). Intertemporal effects of consumption and their implications for demand elasticity estimates. *Quantitative Marketing and Economics*, 4(4), 325–349.

Hasanbeigi, A., & Price, L. (2015). A technical review of emerging technologies for energy and water efficiency and pollution reduction in the textile industry. *Journal of Cleaner Production*, 95, 30–44.

Hausman, J. A. (1978). Specification tests in econometrics. *Econometrica*, 46(6), 1251–1271.

Hayes, S. G., & Venkatraman, P. (eds.). (2016). *Materials and Technology for Sportswear and Performance Apparel*. Boca Raton: CRC Press.

Haynes, J. (2003). Tracing connections between comparative politics and globalisation. *Third World Quarterly*, 24(6), 1029–1047.

HBO, (25 April2015) LasEt week tonight with John Oliver, Season 2, Episode 11, "Fashion"

He, Z., Li, G., Chen, J., Huang, Y., An, T., & Zhang, C. (2015). Pollution characteristics and health risk assessment of volatile organic compounds emitted from different plastic solid waste recycling workshops. *Environment International*, 77, 85–94.

Hedrick, J. B. (1995). The global rare: Earth cycle. *Journal of Alloys and Compounds*, 225(1–2), 609–618.

Hessel, C., Allegre, C., Maisseu, M., Charbit, F., & Moulin, P. (2007). Guidelines and legislation for dye house effluents. *Journal of Environmental Management*, 83(2), 171–180.

Hidrue, M. K., Parsons, G. R., Kempton, W., & Gardner, M. P. (2011). Willingness to pay for electric vehicles and their attributes. *Resource and Energy Economics*, 33(3), 686–705.

Hill, S. E., Rodeheffer, C. D., Griskevicius, V., Durante, K., & White, A. E. (2012). Boosting beauty in an economic decline: Mating, spending, and the lipstick effect. *Journal of Personality and Social Psychology*, 103(2), 275.

Hille, E. (2018). Pollution havens: International empirical evidence using a shadow price measure of climate policy stringency. *Empirical Economics*, 54(3), 1137–1171.

Hollis, N. (2008). *The Global Brand: How to Create and Develop Lasting Brand Value in the World Market*. London: Palgrave Macmillan.

Holkar, C. R., Jadhav, A. J., Pinjari, D. V., Mahamuni, N. M., & Pandit, A. B. (2016). A critical review on textile wastewater treatments: Possible approaches. *Journal of Environmental Management*, 182, 351–366.

Holm, O. (2006). Integrated marketing communication: From tactics to strategy. *Corporate Communications: An International Journal*, (11), 1, 23–33.

Holzemer, R. (2017). *DRIES*. Germany: Reiner Holzemer Films.

Hossain, L., Sarker, S. K., & Khan, M. S. (2018). Evaluation of present and future wastewater impacts of textile dyeing industries in Bangladesh. *Environmental Development*, 26, 23–33.

Houck, O. A. (2002). *The Clean Water Act TMDL Program: Law, Policy, and Implementation*. 2nd Edition. Washington, DC: Environmental Law Institute.

Howard, C. S. (1933). Determination of total dissolved solids in water analysis. *Industrial & Engineering Chemistry Analytical Edition*, 5(1), 4–6.

Howe, P., & Teufel, B. (2014). Native advertising and digital natives: The effects of age and advertisement format on news website credibility judgments. *International Symposium on Online Journalism Journal*, 4(1), 78–90.

Hrdlička, J., & Dlouhý, T. (2019). Full-scale evaluation of SO_2 capture increase for semi-dry FGD technology. *Journal of the Energy Institute*, 92(5), 1399–1405.

Hu, Z. H., Li, Q., Chen, X. J., & Wang, Y. F. (2014). Sustainable rent-based closed-loop supply chain for fashion products. *Sustainability*, 6(10), 7063–7088.

Hwang, I. (2009). A study on characteristics of anti-Hallyu groups and the relationship between anti-Hallyu propensity and the purchase intention for Korean products in China. *Advertising Studies*, 82, 201–231.

Hussain, T., Ashraf, M., Rasheed, A., Ahmad, S., & Ali, Z. (2016). *Textile Engineering: An Introduction*. Walter de Gruyter GmbH & Co KG.

Ibrahim, N. A., Moneim, N. M. A., Halim, E. A., & Hosni, M. M. (2008). Pollution prevention of cotton: Cone reactive dyeing. *Journal of Cleaner Production*, 16(12), 1321–1326.

Inglehart, R. (1995). Public support for environmental protection: Objective problems and subjective values in 43 societies. *Political Science & Politics*, 28(1), 57–72.

Jacometti, V. (2019). Circular economy and waste in the fashion industry. *Laws*, 8(4), 27–40.

James, W. E., Ray, D. J., & Minor, P. J. (2002). Indonesia's Textile and Apparel Industry: Meeting the Challenges of the Changing International Trade Environment. Technical report of the US Agency for International Development and ECG. Indonesia: Jakarta.

Jeacle, I. (2015). Fast fashion: Calculative technologies and the governance of everyday dress. *European Accounting Review*, 24(2), 305–328.

Jebli, M. B., Youssef, S. B., & Ozturk, I. (2016). Testing environmental Kuznets curve hypothesis: The role of renewable and non-renewable energy consumption and trade in OECD countries. *Ecological Indicators*, 60, 824–831.

Jin, B. (2020). Back to the normal: The end of history for hyper growth. In Jin, B. (ed) *China's Path of Industrialization* (pp. 69–101).Singapore: Springer.

Jin, B., & Cedrola, E. (2016). Overview of fashion brand internationalization: Theories and trends. In Jin, B. & Cedrola, E. (eds). *Fashion Brand Internationalization* (pp. 1–30). New York, Palgrave Pivot.

Jin, D. Y. & Yoon, K. (2016). The social mediascape of transnational Korean pop culture: Hallyu 2.0 as spreadable media practice. *New Media & Society* 18(7), 1277–1292.

Johnson, B., & Villumsen, G. (2018). Environmental aspects of natural resource intensive development: The case of agriculture. *Innovation and Development*, 8(1), 167–188.

Jong-Wha, L. E. E., & Wie, D. (2017). Wage structure and gender earnings differentials in China and India. *World Development*, 97, 313–329.

Jorgenson, A. K. (2006). Unequal ecological exchange and environmental degradation: A theoretical proposition and cross-national study of deforestation, 1990–2000. *Rural Sociology*, 71(4), 685–712.

Jorgenson, A. K. (2007). Does foreign investment harm the air we breathe and the water we drink? A cross-national study of carbon dioxide emissions and organic water pollution in less-developed countries, 1975 to 2000. *Organization & Environment*, 20(2), 137–156.

Jorgenson, A. K. (2009). Political-economic integration, industrial pollution and human health: A panel study of less-developed countries, 1980–2000. *International Sociology*, 24(1), 115–143.

Jorgenson, A. K. (2012). The sociology of ecologically unequal exchange and carbon dioxide emissions, 1960–2005. *Social Science Research*, 41(2), 242–252.

Jorgenson, A. K. (2016). The sociology of ecologically unequal exchange, foreign investment dependence and environmental load displacement: Summary of the literature and implications for sustainability. *Journal of Political Ecology*, 23(1), 334–349.

Joo Park, E., Young Kim, E., & Cardona Forney, J. (2006). A structural model of fashion-oriented impulse buying behavior. *Journal of Fashion Marketing and Management: An International Journal*, 10(4), 433–446.

Joung, H. M. (2014). Fast-fashion consumers' post-purchase behaviours. *International Journal of Retail and Distribution Management*, 42(8), 688–697.

Joy, A., Sherry Jr, J. F., Venkatesh, A., Wang, J., & Chan, R. (2012). Fast fashion, sustainability, and the ethical appeal of luxury brands. *Fashion Theory*, 16(3), 273–295.

Kaneva, N. (2011). Nation branding: Toward an agenda for critical research. *International Journal of Communication* 5, 117–141.

Kant, R. (2012). Textile dyeing industry an environmental hazard. *Natural Science*, 4(1), 22–26.

Kapoor, A., & Khare, A. K. (2019). The Afterlife of discarded woollens: Who is recycling my clothes?. *International Journal of Management*, 10(5), 84–98.

Karr, J. R., & Yoder, C. O. (2004). Biological assessment and criteria improve total maximum daily load decision making. *Journal of Environmental Engineering*, 130(6), 594–604.

Keele, L., & Kelly, N. J. (2006). Dynamic models for dynamic theories: The ins and outs of lagged dependent variables. *Political Analysis*, 14(2), 186–205.

Keller, K.L. (2003). *Strategic Brand Management: Building, Measuring, and Managing Brand Equity*. Upper Saddle River, NJ: Prentice Hall.

Keller, W., & Levinson, A. (2002). Pollution abatement costs and foreign direct investment inflows to US states. *Review of Economics and Statistics*, 84(4), 691–703.

Keller, W., & Yeaple, S. R. (2009). Multinational enterprises, international trade, and productivity growth: Firm–level evidence from the United States. *The Review of Economics and Statistics*, 91(4), 821–831.

Kelting, K., & Rice, D. H. (2013). Should we hire David Beckham to endorse our brand? Contextual interference and consumer memory for brands in a celebrity's endorsement portfolio. *Psychology & Marketing*, 30(7), 602–613.

Kentor, J. & Grimes, P. (2006). Foreign investment dependence and the environment: A global perspective, in A. K. Jorgenson & Kick, E. (eds) *Globalization and the Environment*: 61–78. Leiden: Brill Publishing.

Kerr, J. & Landry, J. (2017). *Pulse of the Fashion Industry*. Copenhagen, Denmark: Global Fashion Agenda; Boston, MA, USA: The Boston Consulting Group, 2017. available online: http://globalfashionagenda.com/wp–content/

Keynes, J. M. (1936). *The General Theory of Employment, Interest, and Money*. London: Palgrave Macmillan.

Khan, M. S., Ahmed, S., Evans, A. E., & Chadwick, M. (2009). Methodology for performance analysis of textile effluent treatment plants in Bangladesh. *Chemical Engineering Research Bulletin*, 13(2), 61–66.

Khurana, K., & Ricchetti, M. (2016). Two decades of sustainable supply chain management in the fashion business, an appraisal. *Journal of Fashion Marketing and Management*, 20(2), 89–104.

Kilduff, P., & Chi, T. (2006). Longitudinal patterns of comparative advantage in the textile complex-Part 2: Sectoral perspectives. *Journal of Fashion Marketing and Management: An International Journal*, 10(2), 150–168.

Kim, H. S. (2005). Consumer profiles of apparel product involvement and values. *Journal of Fashion Marketing and Management: An International Journal*, 9(2), 207–220.

Kim, H. S. (2017). An analysis of a strategy for the activation of Korean wave K-fashion. *Journal of the Korea Fashion and Costume Design Association*, 19(3), 175–192.

Kim, J. E., Kim, H. S., Choi, H. S., & Lee, K. M. (2013). Fashion market analysis and consumer research for expansion of Korean wave fashion into the Singapore market. *Fashion & Textile Research Journal*, 15(5), 797–807.

Kim, S. Y. (2019a). Beauty and the waste: Fashioning idols and the ethics of recycling in Korean pop music videos. *Fashion Theory*, 1–21. doi: 10.1080/1362704X.2019.1581001.

Kim, Y. (Ed.). (2019b). *South Korean Popular Culture and North Korea*. London: Routledge.

Kirwan, G. H., Fullwood, C., & Rooney, B. (2018). Risk factors for social networking site scam victimization among Malaysian students. *Cyberpsychology, Behavior, and Social Networking*, 21(2), 123–128.

Klein, N. (1999). *No Logo*. Toronto, ON: Knopf Canada.

Kleinert, J. (2001). The time pattern of the internationalization of production. *German Economic Review*, 2(1), 79–98.

Klepp, I. G., Buck, M., Laitala, K., & Kjeldsberg, M. (2016). What's the problem? Odor-control and the smell of sweat in sportswear. *Fashion Practice*, 8(2), 296–317.

Kliatchko, J. (2005). Towards a new definition of integrated marketing communications (IMC). *International Journal of Advertising*, 24(1), 7–34.
Koh, J. (2005). Alkali-hydrolysis kinetics of alkali-clearable azo disperse dyes containing a fluorosulphonyl group and their fastness properties on PET/cotton blends. *Dyes and Pigments*, 64(1), 17–23.
Kojima, K. (2000). The "flying geese" model of Asian economic development: Origin, theoretical extensions, and regional policy implications. *Journal of Asian Economics*, 11(4), 375–401.
Kollmuss, A. & Agyeman, J. (2002). Mind the gap: Why do people act environmentally and what are the barriers to pro-environmental behavior? *Environmental Education Research*, 8(3), 239–260.
Korber, S., & Reagan, C. (2014). Athleisure trend spells death of denim. *CNBC*, available at: https://www.cnbc.com/2014/08/08/athleisure–trend–spells–death–of–denim.html.
Korhonen, P. (1994). The theory of the flying geese pattern of development and its interpretations. *Journal of Peace Research*, 31(1), 93–108.
Köksal, D., Strähle, J., Müller, M., & Freise, M. (2017). Social sustainable supply chain management in the textile and apparel industry: A literature review. *Sustainability*, 9(1), 100–132.
Krause, E. L. (2018). *Tight knit: Global Families and the Social Life of Fast Fashion*. Chicago, IL: University of Chicago Press.
Kumar, N. (2007). *Private Label Strategy: How to Meet the Store Brand Challenge*. Cambridge, MA: Harvard Business Review Press.
Kunz, G. I., Karpova, E., & Garner, M. B. (2016). *Going Global: The Textile and Apparel Industry*. New York: Fairchild Books.
Lambin, E. F., & Meyfroidt, P. (2011). Global land use change, economic globalization, and the looming land scarcity. *Proceedings of the National Academy of Sciences*, 108(9), 3465–3472.
Lane, J. E. (2017). South, South East and East Asia: Economic miracle but environmental disaster. *Sustainability in Environment*, 2(1), 1–19.
Lane, B. W., Dumortier, J., Carley, S., Siddiki, S., Clark-Sutton, K., & Graham, J. D. (2018). All plug-in electric vehicles are not the same: Predictors of preference for a plug-in hybrid versus a battery-electric vehicle. *Transportation Research Part D: Transport and Environment*, 65, 1–13.
Lane Keller, K. (2001). Mastering the marketing communications mix: Micro and macro perspectives on integrated marketing communication programs. *Journal of Marketing Management*, 17(7–8), 819–847.
Langstaff, A., & Onono, E. (2018). *Fashion's Dirty Secrets*. United Kingdom: Hello Halo Productions.
Lea Wickett, J., Gaskill, L. R., & Damhorst, M. L. (1999). Apparel retail product development: Model testing and expansion. *Clothing and Textiles Research Journal*, 17(1), 21–35.
Lee, C. G. (2009). Foreign direct investment, pollution and economic growth: Evidence from Malaysia. *Applied Economics*, 41(13), 1709–1716.
Leiter, J., Schulman, M. D., & Zingraff, R. (1991). *Hanging by a Thread: Social Change in Southern Textiles*. Ithaka, NY: Cornell University Press.

Leung, K., Bhagat, R., Buchan, N., Erez, M., & Gibson, C. (2005) "Culture and international business: Recent advances and their implications for future research," *Journal of International Business Studies*, 36(4): 357–378.

Levi, S. (January 25, 2018). No, you don't have to wash those jeans – really! levis-trauss.com, available at: https://www.levistrauss.com/2018/01/25/no–dont–wash–jeans–really/.

Levinson, A., & Taylor, M. S. (2008). Unmasking the pollution haven effect. *International Economic Review*, 49(1), 223–254.

Linder, M., & Williander, M. (2017). Circular business model innovation: Inherent uncertainties. *Business Strategy and the Environment*, 26(2), 182–196.

Lipson, S. M., Stewart, S., & Griffiths, S. (2020). Athleisure: A qualitative investigation of a multi-billion-dollar clothing trend. *Body Image*, 32, 5–13.

Liu, S. & Choi, T. (2009). Consumer attitudes towards brand extensions of designer-labels and mass-market labels in Hong Kong. *Journal of Fashion Marketing and Management*, 13(4). 527–540.

Lock, A., & Harris, P. (1996). "Political Marketing – Vive la Difference!" *European Journal of Marketing*, 30(10/11), 14–24.

Lollo, N., & O'Rourke, D. (2020, August 3). Measurement without clear incentives to improve: The impacts of the Higg Facility Environmental Module (FEM) on apparel factory practices and performance. doi: 10.31235/osf.io/g67d8

Love, J. H. (2003). Technology sourcing versus technology exploitation: An analysis of US foreign direct investment flows. *Applied Economics*, 35(15), 1667–1678.

Lu, S. (2012). China takes all? An empirical study on the impacts of quota elimination on world clothing trade from 2000 to 2009. *Journal of Fashion Marketing and Management* 16(3), 306–326.

Lucas, R. E., Wheeler, D., & Hettige, H. (1992). *Economic Development, Environmental Regulation, and the International Migration of Toxic Industrial Pollution, 1960–88* (Vol. 1062). Washington, DC: World Bank Publications.

Luken, R. A., Johnson, F. R., & Kibler, V. (1992). Benefits and costs of pulp and paper effluent controls under the Clean Water Act. *Water Resources Research*, 28(3), 665–674.

Lund, C. (2015). Selling through the senses: Sensory appeals in the fashion retail environment. *Fashion Practice*, 7(1), 9–30.

Lv, F., Yao, D., Wang, Y., Wang, C., Zhu, P., & Hong, Y. (2015). Recycling of waste nylon 6/spandex blended fabrics by melt processing. *Composites Part B: Engineering*, 77, 232–237.

Lynch, D., McGuire, C. J., & Smith, J. A. (2020). Assessing the US sulfur reduction programme in Massachusetts from an environmental justice framework: Is there evidence of disproportionality?. *Journal of Environmental Economics and Policy*, 9(1), 97–110.

Lynch, A., & Strauss, M. (2007). *Changing Fashion: A Critical Introduction to Trend Analysis and Cultural Meaning*. Oxford, UK: Berg.

MacDonald, S., Pan, S., Somwaru, A., & Tuan, F. (2010). China's role in world cotton and textile markets: A joint computable general equilibrium/partial equilibrium approach. *Applied Economics*, 42(7), 875–885.

Machado, M. A. D., de Almeida, S. O., Bollick, L. C., & Bragagnolo, G. (2019). Second-hand fashion market: consumer role in circular economy. *Journal of Fashion Marketing and Management: An International Journal*, 23(3), 382–395.

MacKellar, F. L., Lutz, W., Prinz, C., & Goujon, A. (1995). Population, households, and CO_2 emissions. *Population and Development Review*, 21(4), 849–865.

Madsen, D., & Slåtten, K. (2013). The role of the management fashion arena in the cross-national diffusion of management concepts: The case of the balanced scorecard in the Scandinavian countries. *Administrative Sciences*, 3(3), 110–142.

Malhotra, S. P. K., & Mandal, T. K. (2019). Zinc oxide nanostructure and its application as agricultural and industrial material. In Kumar, V. et al. (eds.) *Contaminants in Agriculture and Environment: Health Risks and Remediation*, 1, 216–223. India: Agro Environ Media.

Malik, N., Maan, A. A., Pasha, T. S., Akhtar, S., & Ali, T. (2010). Role of hazard control measures in occupational health and safety in the textile industry of Pakistan. *Pakistan Journal of Agricultural Science*, 47(1), 72–76.

Mallampally, P., & Sauvant, K. P. (1999). Foreign direct investment in developing countries. *Finance and Development*, 36(1), 34–37.

Manchiraju, S., & Sadachar, A. (2014). Personal values and ethical fashion consumption. *Journal of Fashion Marketing and Management*, 18(3), 357–374.

Mani, M., & Wheeler, D. (1998). In search of pollution havens? Dirty industry in the world economy, 1960 to 1995. *The Journal of Environment & Development*, 7(3), 215–247.

Mansson, M. (2016). Sweden: The world´ s most sustainable country: Political statements and goals for a sustainable society. *Earth Common Journal*, 6(1), 16–22.

Mantzavinos, C. (2004). *Individuals, Institutions, and Markets*. Cambridge, UK: Cambridge University Press.

Marin, D. (1992). Is the export-led growth hypothesis valid for industrialized countries?. *The Review of Economics and Statistics*, 678–688.

Markham, S., & Cangelosi, J. (1999). An international study of unisex and "same-name" fragrance brands. *Journal of Product & Brand Management*, 8(5), 387–401.

Marsh, D., & Sharman, J. C. (2009). Policy diffusion and policy transfer. *Policy Studies*, 30(3), 269–288.

Martin, M. F. (2007). *US clothing and textile trade with China and the world: Trends since the end of quotas. Congressional Research Services Report # RL34106*. Washington, DC: Library of Congress. available at: https://fas.org/sgp/crs/row/RL34106.pdf

Martin, R. H. (1998). *American Ingenuity: Sportswear, 1930s–1970s*. New York: Metropolitan Museum of Art.

Martin, T. (2019). Fashion law needs custom tailored protection for designs. *University of Baltimore Law Review*, 48(3), 453–475.

Maryan, A. S., Montazer, M., & Damerchely, R. (2015). Discoloration of denim garment with color free effluent using montmorillonite based nano clay and enzymes: Nano bio-treatment on denim garment. *Journal of Cleaner Production*, 91, 208–215.

Maurice, M. & Hermann, L. (2017). *The World According to H&M*. France: Premières Lignes Télévision, Java Films.

McColl, J., & Moore, C. (2011). An exploration of fashion retailer own brand strategies. *Journal of Fashion Marketing and Management: An International Journal*, 15(1), 91–107.

McGoldrick, P. J. (1998). Spatial and temporal shifts in the development of international retail images. *Journal of Business Research*, 42(2), 189–196.

McFarlane, A., & Samsioe, E. (2020). # 50+ fashion Instagram influencers: Cognitive age and aesthetic digital labours. *Journal of Fashion Marketing and Management: An International Journal*. doi: 10.1108/JFMM-08-2019-0177

McKelvey, F. (2019). Cranks, clickbait and cons: On the acceptable use of political engagement platforms. *Internet Policy Review*, 8(4), 1–27.

McQueen, R. H., & Vaezafshar, S. (2019). Odor in textiles: A review of evaluation methods, fabric characteristics, and odor control technologies. *Textile Research Journal*, doi: 10.1177/0040517519883952.

Merino, F. (2004). Firms' productivity and internationalization: A statistical dominance test. *Applied Economics Letters*, 11(13), 851–854.

Mihm, B. (2010). Fast fashion in a flat world: Global sourcing strategies. *International Business & Economics Research Journal (IBER)*, 9(6), 55–64.

Mikic, M., Adhikari, R., & Yamamoto, Y. (2008). *Unveiling Protectionism: Regional Responses to Remaining Barriers in the Textile and Clothing Trade*. New York: United Nations ESCAP,

Miller, D., & Woodward, S. (2012). *Blue Jeans: The Art of the Ordinary*. Berkley, CA: University of California Press.

Millimet, D. L., & Roy, J. (2016). Empirical tests of the pollution haven hypothesis when environmental regulation is endogenous. *Journal of Applied Econometrics*, 31(4), 652–677.

Minchin, T. J. (2012). *Empty Mills: The Fight Against Imports and the Decline of the US Textile Industry*. Lanham, MD: Rowman & Littlefield Publishers.

Minot, N. & Daniels, W. L. (2005). Impact of global cotton markets on rural poverty in Benin. *Agricultural Economics*, 33(3), 453–466.

Miroux, A. & Sauvant, K. P. (2005). *TNCs and the Removal of Textiles and Clothing Quotas. UNCTAD Current Studies on FDI and Development*. Geneva, Switzerland: United Nations.

Mitchell, R. B., Andonova, L. B., Axelrod, M., Balsiger, J., Bernauer, T., Green, J. F., Rakhyun E.K., & Morin, J. F. (2020). What we know (and could know) about international environmental agreements. *Global Environmental Politics*, 20(1), 103–121.

Mlachila, M. & Yang, Y. (2004). The end of textiles quotas: A case study of the impact on Bangladesh. Working Paper 04/108. Washington, DC: International Monetary Fund.

Mo, Z. (2015). Internationalization process of fast fashion retailers: Evidence of H&M and Zara. *International Journal of Business and Management*, 10(3), 217–236.

Mohan, S., Muralimohan, N., Vidhya, K., & Sivakumar, C. T. (2017). A case study on-textile industrial process, characterization and impacts of textile effluent. *Indian Journal of Scientific Research*, 17(1), 80–84.

Molla, A. H. & Khan, H. I. (2018). Detoxification of textile effluent by fungal treatment and its performance in agronomic usages. *Environmental Science and Pollution Research*, 25(11), 10820–10828.

Moore, S. B., & Ausley, L. W. (2004). Systems thinking and green chemistry in the textile industry: Concepts, technologies and benefits. *Journal of Cleaner Production*, 12(6), 585–601.

Moore, C. M., Fernie, J., & Burt, S. (2000). Brands without boundaries-the internationalisation of the designer retailer's brand. *European Journal of Marketing*, 34(8), 919–937.

Moorhouse, D. & Moorhouse, D. (2017). Sustainable design: Circular economy in fashion and textiles, *The Design Journal*, 20, 1948–1959.

Moran, D. D., Lenzen, M., Kanemoto, K., & Geschke, A. (2013). Does ecologically unequal exchange occur?. *Ecological Economics*, 89, 177–186.

Morgan, L. R., & Birtwistle, G. (2009). An investigation of young fashion consumers' disposal habits. *International Journal of Consumer Studies*, 33(2), 190–198.

Morris, M., Plank, L., & Staritz, C. (2016). Regionalism, end markets and ownership matter: Shifting dynamics in the apparel export industry in Sub Saharan Africa. *Environment and Planning A: Economy and Space*, 48(7), 1244–1265.

Tachizawa, M. E., & Yew Wong, C. (2014). Towards a theory of multi-tier sustainable supply chains: A systematic literature review. *Supply Chain Management: An International Journal*, 19(5/6), 643–663.

Müller, J., & Christandl, F. (2019). Content is king-But who is the king of kings? The effect of content marketing, sponsored content & user-generated content on brand responses. *Computers in Human Behavior*, 96, 46–55.

Muruganantham, G., & Bhakat, R. S. (2013). A review of impulse buying behavior. *International Journal of Marketing Studies*, 5(3), 149.

Myzelev, A. (Ed.). (2017). *Fashion, Interior Design and the Contours of Modern Identity*. London: Routledge.

Nagubadi, R. V., & Zhang, D. (2008). Foreign direct investment outflows in the forest products industry: The case of the United States and Japan. *International Forestry Review*, 10(4), 632–640.

Narwal, K. P., & Jindal, S. (2015). The impact of corporate governance on the profitability: An empirical study of Indian textile industry. *International Journal of Research in Management, Science & Technology*, 3(2), 81–85.

Neary, J. P. (2009). Trade costs and foreign direct investment. *International Review of Economics & Finance*, 18(2), 207–218.

Nehf, J. P. (2018). Misleading and unfair advertising. In Howells, G., Ramsey, I., & Wilhelmsson T.. (eds). *Handbook of Research on International Consumer Law*, 2nd Edition. Cheltenham, UK: Edward Elgar Publishing.

Netchaeva, E., & Rees, M. (2016). Strategically stunning: The professional motivations behind the lipstick effect. *Psychological Science*, 27(8), 1157–1168.

Nguyen, D. H., de Leeuw, S., & Dullaert, W. E. (2018). Consumer behaviour and order fulfilment in online retailing: A systematic review. *International Journal of Management Reviews*, 20(2), 255–276.

Nidumolu, R., Ellison, J., Whalen, J., & Billman, E. (2014). The collaboration imperative. *Harvard Business Review*, 92(4), 76–84.

Niinimäki, K., & Hassi, L. (2011). Emerging design strategies in sustainable production and consumption of textiles and clothing. *Journal of Cleaner Production*, 19(16), 1876–1883.

Niinimäki, K., Peters, G., Dahlbo, H., Perry, P., Rissanen, T., & Gwilt, A. (2020). The environmental price of fast fashion. *Nature Reviews Earth & Environment*, 1, 189–200.

Nimkar, U. (2018). Sustainable chemistry: A solution to the textile industry in a developing world. *Current Opinion in Green and Sustainable Chemistry*, 9, 13–17.

Nnorom, I. C., & Osibanjo, O. (2008). Overview of electronic waste (e-waste) management practices and legislations, and their poor applications in the developing countries. *Resources, Conservation and Recycling*, 52(6), 843–858.

Norberg-Bohm, V., & Rossi, M. (1998). The power of incrementalism: Environmental regulation and technological change in pulp and paper bleaching in the US. *Technology Analysis & Strategic Management*, 10(2), 225–245.

Nucamendi-Guillén, S., Moreno, M. A., & Mendoza, A. (2018). A methodology for increasing revenue in fashion retail industry. *International Journal of Retail & Distribution Management*, 46(8), 726–743.

Nueno, J. L., & Quelch, J. A. (1998). The mass marketing of luxury. *Business Horizons*, 41(6), 61–61.

O'Cass, A., & Siahtiri, V. (2013). In search of status through brands from Western and Asian origins: Examining the changing face of fashion clothing consumption in Chinese young adults. *Journal of Retailing and Consumer Services*, 20(6), 505–515.

Oh, S. (2016). Hallyu (Korean Wave) as Korea's cultural public diplomacy in China and Japan. in Ayhan, K. (ed.) *Korea's Public Diplomacy*. Korea: Seoul National University Press, Seoul.

Oketola, A. A., & Osibanjo, O. (2007). Estimating sectoral pollution load in Lagos by industrial pollution projection system (IPPS). *Science of the Total Environment*, 377(2–3), 125–141.

Okonkwo, U. (2007). *Luxury Fashion Branding: Trends, Tactics, Techniques*. New York: Palgrave Macmillan.

Oosterhuis D. M., Weir B. L. (2010) Foliar fertilization of cotton. In: Stewart, J.M., Oosterhuis, D.M., Heitholt, J.J., Mauney, J.R. (eds) *Physiology of Cotton*. Dordrecht: Springer.

Ozdamar Ertekin, Z., & Atik, D. (2015). Sustainable markets: Motivating factors, barriers, and remedies for mobilization of slow fashion. *Journal of Macromarketing*, 35(1), 53–69.

Pablo-Romero, M. D. P., & Sánchez-Braza, A. (2017). The changing of the relationships between carbon footprints and final demand: Panel data evidence for 40 major countries. *Energy Economics*, 61, 8–20.

Pal, H., Chatterjee, K. N., & Sharma, D. (2017). Water footprint of denim industry. In Muthu, S.S. (ed) *Sustainability in Denim* (pp. 111–123). Duxford, UK: Woodhead Publishing.

Pal, R., & Gander, J. (2018). Modelling environmental value: An examination of sustainable business models within the fashion industry. *Journal of Cleaner Production*, 184, 251–263.

Palmer, A. (2001). *Couture & Commerce: The Transatlantic Fashion Trade in the 1950s*. Toronto: ON, UBC Press.

Palumbo, F., & Herbig, P. (2000). The multicultural context of brand loyalty. *European Journal of Innovation Management*, 3(3), 116–125.

Pan, J., Chu, C., Zhao, X., Cui, Y., & Voituriez, T. (2008). Global cotton and textile product chains: Identifying challenges and opportunities for China through a global commodity chain sustainability analysis. *International Institute for Sustainable Development (IISD)*, Winnipeg, Manitoba.

Panke, D. (2020). Inside international environmental organizations: Negotiating the greening of international politics. *Cambridge Review of International Affairs*, 33(3), 365–384.

Paras, M. K., Ekwall, D., Pal, R., Curteza, A., Chen, Y., & Wang, L. (2018). An exploratory study of Swedish charities to develop a model for the reuse-based clothing value chain. *Sustainability*, 10(4), 1176.

Park, J. (2011). The aesthetic style of Korean singers in Japan: A review of Hallyu from the perspective of fashion. *International Journal of Business and Social Science*, 2(19), 23–34.

Park, H. J., & Lin, L. M. (2018). Exploring attitude-behavior gap in sustainable consumption: Comparison of recycled and upcycled fashion products. *Journal of Business Research*. doi: 10.1016/j.jbusres.2018.08.025

Parment, A. (2013). Generation Y vs. Baby Boomers: Shopping behavior, buyer involvement and implications for retailing. *Journal of Retailing and Consumer Services*, 20(2), 189–199.

Partzsch, L. (2015). The power of celebrities in global politics. *Celebrity Studies*, 6(2), 178–191.

Patt, A., Aplyn, D., Weyrich, P., & van Vliet, O. (2019). Availability of private charging infrastructure influences readiness to buy electric cars. *Transportation Research Part A: Policy and Practice*, 125, 1–7.

Pattanayak, A. K. (2020). Sustainability in fabric manufacturing. In Nayak, R. (ed) *Sustainable Technologies for Fashion and Textiles* (pp. 57–72). Cambridge, UK: Woodhead Publishing.

Pattanayak, S. K., Wunder, S., & Ferraro, P. J. (2010). Show me the money: Do payments supply environmental services in developing countries?. *Review of Environmental Economics and Policy*, 4(2), 254–274.

Paul, J. (2015). Masstige marketing redefined and mapped: Introducing a pyramid model and MMS measure. *Marketing Intelligence & Planning*, 33(5), 691–706.

Paul, B. K., & De, S. (2000). Arsenic poisoning in Bangladesh: A geographic analysis. *Journal of the American Water Resources Association*, 36(4), 799–809.

Pengnate, S. F. (2019). Shocking secret you won't believe! Emotional arousal in clickbait headlines. *Online Information Review*, 43(7), 1136–1150.

Pensupa, N., Leu, S. Y., Hu, Y., Du, C., Liu, H., Jing, H., & Lin, C. S. K. (2017). Recent trends in sustainable textile waste recycling methods: Current situation and future prospects. In Lin, C. (ed.) *Chemistry and Chemical Technologies in Waste Valorization* (pp. 189–228). Cham, Switzerland: Springer.

Pentecost, R., & Andrews, L. (2010). Fashion retailing and the bottom line: The effects of generational cohorts, gender, fashion fanship, attitudes and impulse buying on fashion expenditure. *Journal of Retailing and Consumer Services*, 17(1), 43–52.

Perman, R., Ma, Y., McGilvray, J., & Common, M. (2003). *Natural Resource and Environmental Economics*, 3rd edition, Harlow, UK: Pearson Education Limited.

Perraton, J. (2019). The scope and implications of globalisation. In Michie, J. (ed) *The Handbook of Globalisation*, 3rd Edition. Cheltenham, UK: Edward Elgar Publishing.

Perraton, J., Goldblatt, D., Held, D., & McGrew, A. (1997). The globalisation of economic activity. *New Political Economy*, 2(2), 257–277.

Perry, P., Wood, S. (2019) "Exploring the international fashion supply Chain and corporate social responsibility: Cost, responsiveness and ethical implications". In Fernie, J. and Sparks, L. (eds.) *(2019) Logistics and Retail Management*, 5th Edition, London: Kogan Page.

Pettinger, L. (2004). Brand culture and branded workers: Service work and aesthetic labour in fashion retail. *Consumption Markets & Culture*, 7(2), 165–184.

Petty, R. D. (1997). Advertising law in the United States and European Union. *Journal of Public Policy & Marketing*, 16(1), 2–13.

Phau, I., & Lo, C. C. (2004). Profiling fashion innovators. *Journal of Fashion Marketing and Management: An International Journal*, 8(4), 399–411.

Piketty, T. (2006). The Kuznets curve: Yesterday and tomorrow. in Banerjee, A. V., Benabou, R. & Mookherjee, D. (eds) *Understanding Poverty*, Oxford, UK: Oxford University Press, pp. 63–72.

Plank, L., Rossi, A., & Staritz, C. (2012). Workers and social upgrading in" fast fashion": The case of the apparel industry in Morocco and Romania. Working Paper No. 33. Vienna, Austria: Austrian Foundation for Development Research (ÖFSE). available at: https://www.oefse.at/fileadmin/content/Downloads/Publikationen/Workingpaper/WP33_fast_fashion.pdf

Pledge, R. (December 8, 2018). *A photographer goes missing in China. The New York Times*, available at: https://www.nytimes.com/2018/12/08/opinion/sunday/lu-guang-photographer-missing-china.html

Polegato, R., & Wall, M. (1980). Information seeking by fashion opinion leaders and followers. *Home Economics Research Journal*, 8(5), 327–338.

Pomodoro, S. (2013). Temporary retail in fashion system: an explorative study. *Journal of Fashion Marketing and Management: An International Journal*, 17(3), 341–352.

Ponte, S. & Richey, L.A. (2014). Buying into development? Brand aid forms of cause-related marketing. *Third World Quarterly*, 35(1), 65–87.

Pookulangara, S., & Shephard, A. (2013). Slow fashion movement: Understanding consumer perceptions—An exploratory study. *Journal of Retailing and Consumer Services*, 20(2), 200–206.

Popp, D. (2006). International innovation and diffusion of air pollution control technologies: The effects of NOX and SO_2 regulation in the US, Japan, and Germany. *Journal of Environmental Economics and Management*, 51(1), 46–71.

Potvin, J. (2017). *Giorgio Armani: Empire of the Senses*. London: Routledge.

Prabhu, K. H., & Bhute, A. S. (2012). Plant based natural dyes and mordants: A Review. *Journal of Natural Production and Plant Resources*, 2(6), 649–664.

Prasad, M. P. D., Sridevi, V., Lakshmi, P. K., Swathi, A. (2015). Treatment of pharmaceutical industrial effluent by microbial fuel cell (MFC). *International Journal for Innovative Research in Science and Technology*, 2(1), 241–247

Priest, A. (2005). Uniformity and differentiation in fashion. *International Journal of Clothing Science and Technology*, 17(3/4), 253–263.

Rahman, O., & Gong, M. (2016). Sustainable practices and transformable fashion design-Chinese professional and consumer perspectives. *International Journal of Fashion Design, Technology and Education*, 9(3), 233–247.

Rajamani, S. (2016). Novel industrial wastewater treatment integrated with recovery of water and salt under a zero liquid discharge concept. *Reviews on Environmental Health*, 31(1), 63–66.

Ramamoorthy, S. K., Persson, A., & Skrifvars, M. (2014). Reusing textile waste as reinforcements in composites. *Journal of Applied Polymer Science*, 131(17), 1–16.

Ramondo, N. (2009). Foreign plants and industry productivity: Evidence from Chile. *Scandinavian Journal of Economics*, 111(4), 789–809.

Ransom, B. (2020). *The true cost of colour: The impact of textile dyes on water systems*. fashionrevolution.org, available at: https://www.fashionrevolution.org/the-true-cost-of-colour-the-impact-of-textile-dyes-on-water-systems/

Rantisi, N. M. (2004). The ascendance of New York fashion. *International Journal of Urban and Regional Research*, 28(1), 86–106.

Rashid, A., & Barnes, L. (2017). Country of origin: Reshoring implication in the context of the UK fashion industry. In Vecchi A. (ed). *Reshoring of Manufacturing* (pp. 183–201) Cham, Switzerland: Springer.

Rasli, A. M., Qureshi, M. I., Isah-Chikaji, A., Zaman, K., & Ahmad, M. (2018). New toxics, race to the bottom and revised environmental Kuznets curve: The case of local and global pollutants. *Renewable and Sustainable Energy Reviews*, 81, 3120–3130.

Rastogi, A., Choi, J. K., Hong, T., & Lee, M. (2017). Impact of different LEED versions for green building certification and energy efficiency rating system: A Multifamily Midrise case study. *Applied Energy*, 205, 732–740.

Rathinamoorthy, R. (2018). Consumer's Awareness on Sustainable Fashion (pp. 1–36). In Subramanian, S. (ed). *Sustainable Fashion: Consumer Awareness and Education*, Singapore: Springer.

Reddy, M., Terblanche, N., Pitt, L., & Parent, M. (2009). How far can luxury brands travel? Avoiding the pitfalls of luxury brand extension. *Business Horizons*, 52(2), 187–197.

Reed, D. (2013). *Structural Adjustment, the Environment and Sustainable Development*. London, UK: Routledge.

Remington, C. (18 February, 2020). *France considers labelled apparel ratings*. Ecotextile.com, available at: https://www.ecotextile.com/2020021825710/fashion-retail-news/france-considers-environmental-ratings-on-apparel.html

Remy, N., Speelman, E., & Swartz, S. (2016). Style that's sustainable: A new fast-fashion formula. *McKinsey & Company*, 1–6, available at: https://www.mckinsey.com/business-functions/sustainability/our-insights/style-thats-sustainable-a-new-fast-fashion-formula#

Reyneke, M., Sorokáčová, A., & Pitt, L. (2012). Managing brands in times of economic downturn: How do luxury brands fare?. *Journal of Brand Management*, 19(6), 457–466.

Rice, J. (2007). Ecological unequal exchange: International trade and uneven utilization of environmental space in the world system. *Social Forces*, 85(3): 1369–1392.

Ringquist, E. J. (1993). Does regulation matter?: Evaluating the effects of state air pollution control programs. *The Journal of Politics*, 55(4), 1022–1045.

Rivera-Batiz, F. L., & Rivera-Batiz, L. A. (1990). The effects of direct foreign investment in the presence of increasing returns due to specialization. *Journal of Development Economics*, 34(1–2), 287–307.

Roberts, J. and Cayla, J. (2009). Global branding. in Kotabe, M. and Helsen, K. (Eds.). *The SAGE Handbook of International Marketing*. London, UK: Sage. pp346–360.

Roberts, J. T., & Parks, B. C. (2009). Ecologically unequal exchange, ecological debt, and climate justice: The history and implications of three related ideas for a new social movement. *International Journal of Comparative Sociology*, 50 (3–4), 385–409.

Robinson, B. H. (2009). E-waste: An assessment of global production and environmental impacts. *Science of the Total Environment*, 408(2), 183–191.

Robinson, R. (2008). *Indonesia: e Rise of Capital*. Sheeld, UK: Equinox Press.

Robison, R. (2009). *Indonesia: The Rise of Capital*. Sheffield, UK: Equinox Publishing.

Rodrik, D. (Ed.). (2003). *In Search of Prosperity: Analytic Narratives on Economic Growth*. Princeton, NJ: Princeton University Press.

Rose, M., Böhringer, B., Jolly, M., Fischer, R., & Kaskel, S. (2011). MOF processing by electrospinning for functional textiles. *Advanced Engineering Materials*, 13(4), 356–360.

Rosenthal, E. (January 25, 2007). Can polyester save the world? *The New York Times*. available at: https://www.nytimes.com/2007/01/25/fashion/25pollute.html.

Ross, J., & Harradine, R. (2010). Value brands: cheap or trendy? An investigation into young consumers and supermarket clothing. *Journal of Fashion Marketing and Management: An International Journal*, 14(3), 350–366.

Ross, M. & Morgan, A. (2015). *The True Cost*. United States: Untold Creative, LLC.

Rothman, D. S. (1998). Environmental Kuznets curves—real progress or passing the buck?: A case for consumption-based approaches. *Ecological Economics*, 25(2), 177–194.

Rudra, A., & Chattopadhyay, A. (2018). Environmental quality in India: Application of environmental Kuznets curve and Sustainable Human Development Index. *Environmental Quality Management*, 27(4), 29–38.

Sakamoto, M., Ahmed, T., Begum, S., & Huq, H. (2019). Water pollution and the textile industry in Bangladesh: Flawed corporate practices or restrictive opportunities?. *Sustainability*, 11(7), 1951–1965.

Samanta, K. K., Basak, S., & Chattopadhyay, S. K. (2017). Sustainable dyeing and finishing of textiles using natural ingredients and water-free technologies. In Muthu, S.S. (ed) *Textiles and Clothing Sustainability* (pp. 99–131).Singapore: Springer.

Sandvik, I. M., & Stubbs, W. (2019). Circular fashion supply chain through textile-to-textile recycling. *Journal of Fashion Marketing and Management: An International Journal*, 23(3), 366–381.

Sanghani, R. (2018, October 9). Stacey Dooley investigates: Are your clothes wrecking the planet. bbc.com, available at: https://www.bbc.co.uk/bbcthree/article/5a1a43b5-cbae-4a42-8271-48f53b63bd07.

Schielke, T., & Leudesdorff, M. (2015). Impact of lighting design on brand image for fashion retail stores. *Lighting Research & Technology*, 47(6), 672–692.

Schivinski, B., & Dabrowski, D. (2016). The effect of social media communication on consumer perceptions of brands. *Journal of Marketing Communications*, 22(2), 189–214.

Schlosberg, D. (2009). *Defining Environmental Justice: Theories, Movements, and Nature*. Oxford, UK: Oxford University Press.

Schofer, E. & Hironaka, A. (2005). The effects of world society on environmental protection outcomes. *Social Forces*, 84(1), 25–47.

Schrank, H. L., & Lois Gilmore, D. (1973). Correlates of fashion leadership: Implications for fashion process theory. *Sociological Quarterly*, 14(4), 534–543.

Schröppel, C., & Mariko, N. (2003). The changing interpretation of the flying geese model of economic development. *Japanstudien*, 14(1), 203–236.

Schultz, D. E. (1992). Integrated marketing communications. *Journal of Promotion Management*, 1(1), 99–104.

Schwartz, P. (Ed.). (2008). *Structure and Mechanics of Textile Fibre Assemblies*. Boca Raton, FL: CRC Press.

Sealey, I. J. E., Persinger Jr, W. H., Robarge, K., & Luo, M. (2004). U.S. Patent No. 6, 686,039. Washington, DC: U.S. Patent and Trademark Office.

Shahbaz, M., Dube, S., Ozturk, I., & Jalil, A. (2015a). Testing the environmental Kuznets curve hypothesis in Portugal. *International Journal of Energy Economics and Policy*, 5(2), 475–481.

Shahbaz, M., Solarin, S. A., Sbia, R., & Bibi, S. (2015b). Does energy intensity contribute to CO_2 emissions? A trivariate analysis in selected African countries. *Ecological indicators*, 50, 215–224.

Shan, J., & Sun, F. (1998). On the export-led growth hypothesis: The econometric evidence from China. *Applied Economics*, 30(8), 1055–1065.

Shandra, J. M., Shor, E., & London, B. (2008). Debt, structural adjustment, and organic water pollution: A cross-national analysis. *Organization & Environment*, 21(1), 38–55.

Shaw, D., Hogg, G., Wilson, E., Shiu, E., & Hassan, L. (2006). Fashion victim: The impact of fair trade concerns on clothing choice. *Journal of Strategic Marketing*, 14(4), 427–440.

Shin, S. H., Kim, H. O., & Rim, K. T. (2019). Worker safety in the rare earth elements recycling process from the review of toxicity and issues. *Safety and Health at Work*, 10(4), 409–419.

Shin, E. J., & Koh, A. R. (2020). Korean Genderless Fashion Consumers' Self-image and Identification. *Journal of the Korean Society of Clothing and Textiles*, 44(3), 400–412.

Shirazi, N. S., & Manap, T. A. A. (2005). Export-led growth hypothesis: Further econometric evidence from South Asia. *The Developing Economies*, 43(4), 472–488.

Shishoo, R. (Ed.). (2015). *Textiles for Sportswear*. London: Elsevier.

Shipchandler, Z. E. (1982). Keeping down with the Joneses: Stagflation and buyer behavior. *Business Horizons*, 25(6), 32–38.

Siddiki, S., Dumortier, J., Curley, C., Graham, J. D., Carley, S., & Krause, R. M. (2015). Exploring drivers of innovative technology sdoption intention: The case of plug: In vehicles. *Review of Policy Research*, 32(6), 649–674.

Silva, T. L., Cazetta, A. L., Souza, P. S., Zhang, T., Asefa, T., & Almeida, V. C. (2018). Mesoporous activated carbon fibers synthesized from denim fabric waste: Efficient adsorbents for removal of textile dye from aqueous solutions. *Journal of Cleaner Production*, 171, 482–490.

Simmons, B. A., & Elkins, Z. (2004). The globalization of liberalization: Policy diffusion in the international political economy. *American Political Science Review*, 98(1), 171–189.

Singh, R. L., Singh, P. K., & Singh, R. P. (2015). Enzymatic decolorization and degradation of azo dyes-A review. *International Biodeterioration & Biodegradation*, 104, 21–31.

Singleton, J. (2013). *World Textile Industry*. London: Routledge.

Siuli, A., & Mondal, B. (2017). Treatment of textile effluent by using potential microbes. *International Journal of Engineering Science*, 7(4), 10146–10148.

Smarzynska, B. K., & Wei, S. J. (2001). Pollution havens and foreign direct investment: Dirty secret or popular myth? Working paper No. w8465. Washington, DC: National Bureau of Economic Research.

Smith, A. N., Fischer, E., & Yongjian, C. (2012). How does brand-related user-generated content differ across You Tube, Facebook, and Twitter?. *Journal of Interactive Marketing*, 26(2), 102–113.

Snyder, L. B., Willenborg, B., & Watt, J. (1991). Advertising and cross-cultural convergence in Europe, 1953–89. *European Journal of Communication*, 6(4), 441–468.

Sodhi, N. (2017). New business model for the global fashion industry in disruptive times. *Textile Times*, 15(4), 4–10.

Sodhi, N. (2018). Technology weds fashion: The global convergence. *Textile Times*, 15(7), 48–52.

Soule, P. T., & DeHart, J. L. (1998). Assessing IPAT using production-and consumption-based measures of I. *Social Science Quarterly*, 79(4), 754–765.

Srinivasan, R. (2015). Exploring the impact of social norms and online shopping anxiety in the adoption of online apparel shopping by Indian consumers. *Journal of Internet Commerce*, 14(2), 177–199.

Sriramesh, K., & Vercic, D. (Eds.). (2003). *The Global Public Relations Handbook: Theory, Research, and Practice*. London, UK: Routledge.

Srivastava, R. K., Jozewicz, W., & Singer, C. (2001). SO_2 scrubbing technologies: A review. *Environmental Progress*, 20(4), 219–228.

Statista. (2019). *Global Apparel Market-Statistics and Facts*. statista.com, available at: https://www.statista.com/topics/5091/apparel–market–worldwide/

Stearns, P. N. (2006). *Consumerism in World History: The Global Transformation of Desire*. London, UK: Routledge.

Stearns, P. N. (2009). Consumerism. in Pell, J., & van Staveren, I. (eds). *Handbook of Economics and Ethics*, (pp. 62–76). Cheltenham, UK: Edward Elgar Publishing.

Steenkamp, J. B. (2014). How global brands create firm value: the 4V model. *International Marketing Review*, 31(1), 5–29.

Steenkamp, J. B. E., Batra, R., & Alden, D. L. (2003). How perceived brand globalness creates brand value. *Journal of International Business Studies*, 34(1), 53–65.

Stephen Parker, R., Hermans, C. M., & Schaefer, A. D. (2004). Fashion consciousness of Chinese, Japanese and American teenagers. *Journal of Fashion Marketing and Management: An International Journal*, 8(2), 176–186.

Stern, D. I., Common, M. S., & Barbier, E. B. (1996). Economic growth and environmental degradation: The environmental Kuznets curve and sustainable development. *World Development*, 24(7), 1151–1160.

Stern, P. C. (2000). New environmental theories: Toward a coherent theory of environmentally significant behavior. *Journal of Social Issues*, 56(3), 407–424.

Stiglitz, J. E. (2003). Democratizing the International Monetary Fund and the World Bank: Governance and accountability. *Governance*, 16(1), 111–139.

Stretesky, P. B., & Lynch, M. J. (2009). A cross–national study of the association between per capita carbon dioxide emissions and exports to the United States. *Social Science Research*, 38(1), 239–250.

Strizhakova, Y., Coulter, R. A., & Price, L. (2008). "Branded products as a passport to global citizenship: perspectives from developed and developing countries." *Journal of International Marketing* 16(4), 57–85.

Sudha, M., & Sheena, K. (2017). Impact of influencers in consumer decision process: The fashion industry. *SCMS Journal of Indian Management*, 14(3), 14–30.

Summers, J. O. (1970). The identity of women's clothing fashion opinion leaders. *Journal of Marketing Research*, 7(2), 178–185.

Szirmai, A. (2012). Industrialization as an engine of growth in developing countries, 1950–2005. *Structural Change and Economic Dynamics*, 23(4), 406–420.

Tadesse, M. G., Harpa, R., Chen, Y., Wang, L., Nierstrasz, V., & Loghin, C. (2019). Assessing the comfort of functional fabrics for smart clothing using subjective evaluation. *Journal of Industrial Textiles*, 48(8), 1310–1326.

Taplin, I. M. (2014). Who is to blame?: A re-examination of fast fashion after the 2013 factory disaster in Bangladesh. *Critical Perspectives on International Business*, 10(1–2), 72–83.

Taylor, M. R., Rubin, E. S., & Hounshell, D. A. (2005). Control of SO_2 emissions from power plants: A case of induced technological innovation in the US. *Technological Forecasting and Social Change*, 72(6), 697–718.

Thomas, D. (2019) *Fashionopolis: The Price of Fast Fashion and the Future of Our Clothes*. London: Head of Zeus.

Timmer, M., Los, B., Stehrer, R., & de Vries, G. (2016). *An anatomy of the global trade slowdown based on the WIOD 2016 release (No. GD–162)*. Netherlands: Groningen Growth and Development Centre, University of Groningen.

Tirunillai, S., & Tellis, G. J. (2012). Does chatter really matter? Dynamics of user-generated content and stock performance. *Marketing Science*, 31(2), 198–215.

Tisdell, C. (2001). Globalisation and sustainability: Environmental Kuznets curve and the WTO. *Ecological Economics*, 39(2), 185–196.

To, M.H., Uisan, K., Ok, Y. S., Pleissner, D., & Lin, C. S. K. (2019). Recent trends in green and sustainable chemistry: Rethinking textile waste in a circular economy. *Current Opinion in Green and Sustainable Chemistry*, 20, 1–10.

Todeschini, B. V., Cortimiglia, M. N., Callegaro-de-Menezes, D., & Ghezzi, A. (2017). Innovative and sustainable business models in the fashion industry: Entrepreneurial drivers, opportunities, and challenges. *Business Horizons*, 60(6), 759–770.

Tokatli, N. (2008). Global Sourcing: Insights from the global clothing industry-the case of Zara, a fast fashion Retailer. *Journal of Economic Geography*, 8, 21–38.

Tokatli, N., & Kızılgün, Ö. (2009). From manufacturing garments for ready-to-wear to designing collections for fast fashion: Evidence from Turkey. *Environment and Planning A*, 41(1), 146–162.

Tong, T., & Elimelech, M. (2016). The global rise of zero liquid discharge for wastewater management: Drivers, technologies, and future directions. *Environmental Science & Technology*, 50(13), 6846–6855.

Trumbull, G. (2018). *Consumer Capitalism: Politics, Product Markets and Firm Strategy in France and Germany*. New York: Cornell University Press.

Truong, Y., McColl, R., & Kitchen, P. J. (2009). New luxury brand positioning and the emergence of masstige brands. *Journal of Brand Management*, 16(5–6), 375–382.

Tull, D. M. (2006). China's engagement in Africa: Scope, significance and consequences. *The Journal of Modern African Studies*, 44(3), 459–479.

Tungate, M. (2008). *Fashion Brands: Branding Style from Armani to Zara*. London, UK: Kogan Page Publishers.

Turker, D., & Altuntas, C. (2014). Sustainable supply chain management in the fast fashion industry: An analysis of corporate reports. *European Management Journal*, 32(5), 837–849.

Tüfekci, N., Sivri, N., & Toroz, I. (2007). Pollutants of textile industry wastewater and assessment of its discharge limits by water quality standards. *Turkish Journal of Fisheries and Aquatic Science*, 7(2), 97–103.

UCRF (5 May 2019). *Statement on the 2019 Copenhagen Fashion Summit.* Concernedresearchers.org, available at: http://www.concernedresearchers.org/ucrf-on-2019-copenhagen-fashion-summit/.

Ulrich, P. V., Anderson-Connell, L. J., & Wu, W. (2003). Consumer co-design of apparel for mass customization. *Journal of Fashion Marketing and Management: An International Journal*, 7(4), 398–412.

UNFCCC (10 December 2018). *United Nations Framework Convention on Climate Change. Fashion Industry Charter for Climate Action.* unfccc.int, available at: https://unfccc.int/sites/default/files/resource/Industry%20Charter%20%20Fashion%20and%20Climate%20Action%20–%2022102018.pdf.

Upton, G. B., & Snyder, B. F. (2015). Renewable energy potential and adoption of renewable portfolio standards. *Utilities Policy*, 36, 67–70.

Upton, G. B., & Snyder, B. F. (2017). Funding renewable energy: An analysis of renewable portfolio standards. *Energy Economics*, 66, 205–216.

US EPA. (2019). *Clean Water Act, Section 303(d), Impaired Waters and Total Maximum Daily Loads (TDMLs)*, available at: https://www.epa.gov/tmdl.

Uusivuori, J., & Laaksonen-Craig, S. (2001). Foreign direct investment, exports and exchange rates: The case of forest industries. *Forest Science*, 47(4), 577–586.

Uygur, A. (2017). The future of organic fibres. *European Journal of Sustainable Development Research*, 2(1), 164–172.

Vamvakidis, M. A., & Dodzin, M. S. (1999). Trade and industrialization in developing agricultural economies. Working Paper No. 99–145. Washington, DC: International Monetary Fund.

Van Calster, G. (2015). *EU Waste Law.* Oxford: Oxford University Press.

Van Dijck, J. (2009). Users like you? Theorizing agency in user-generated content. *Media, Culture & Society*, 31(1), 41–58.

Van Gelder, S. (2005). The new imperatives for global branding: Strategy, creativity and leadership. *Journal of Brand Management*, 12(5), 395–404.

Vandevivere, P. C., Bianchi, R., & Verstraete, W. (1998). Treatment and reuse of wastewater from the textile wet-processing industry: Review of emerging technologies. *Journal of Chemical Technology & Biotechnology: International Research in Process, Environmental AND Clean Technology*, 72(4), 289–302.

Vehmas, K., Raudaskoski, A., Heikkilä, P., Harlin, A., & Mensonen, A. (2018). Consumer attitudes and communication in circular fashion. *Journal of Fashion Marketing and Management: An International Journal*, 22(3), 286–300.

Verones, F., Moran, D., Stadler, K., Kanemoto, K., & Wood, R. (2017). Resource footprints and their ecosystem consequences. *Scientific Reports*, 7, article # 40743.

Vreeland, L. I., Tcheng, F., & Perlmutt, B-.J. (2011). *Diana Vreeland: The Eye Has to Travel*. New York, NY: Gloss Studios.
Vultee, F., Burgess, G. S., Frazier, D., & Mesmer, K. (2020). Here's what to know about clickbait: Effects of image, headline and editing on audience attitudes. *Journalism Practice*, 1–18. doi: 10.1080/17512786.2020.1793379.
Wada, K. (1992). The development of tiered inter-firm. *Japanese Yearbook on Business History*, 8, 23–47.
Wagner, U. J., & Timmins, C. D. (2009). Agglomeration effects in foreign direct investment and the pollution haven hypothesis. *Environmental and Resource Economics*, 43(2), 231–256.
Wakelyn, P. (2006). *Cotton Fiber Chemistry and Technology*. Boca Raton: CRC Press.
Wanassi, B., Azzouz, B., & Hassen, M. B. (2016). Value-added waste cotton yarn: Optimization of recycling process and spinning of reclaimed fibers. *Industrial Crops and Products*, 87, 27–32.
Watharkar, A. D., Khandare, R. V., Waghmare, P. R., Jagadale, A. D., Govindwar, S. P., & Jadhav, J. P. (2015). Treatment of textile effluent in a developed phytoreactor with immobilized bacterial augmentation and subsequent toxicity studies on Etheostoma olmstedi fish. *Journal of Hazardous Materials*, 283, 698–704.
Webber, K. (22 March, 2017). *How fast fashion is killing rivers worldwide*. Ecowatch.com, available at: https://www.ecowatch.com/fast–fashion–river-blue–2318389169.html.
Weinzettel, J., Hertwich, E. G., Peters, G. P., Steen–Olsen, K., & Galli, A. (2013). Affluence drives the global displacement of land use. *Global Environmental Change*, 23(2), 433–438.
Wen, D., & Li, M. (2007). China: Hyper–development and environmental crisis. *Socialist Register* 43, 130–145.
Wen, X., Choi, T. M., & Chung, S. H. (2019). Fashion retail supply chain management: A review of operational models. *International Journal of Production Economics*, 207, 34–55.
Wenig, M. M. (1998). How total are total maximum daily loads-legal issues regarding the scope of watershed-based pollution control under the clean water act. *Tulane Environmental Law Journal* 12, 87–118.
Wenting, R., & Frenken, K. (2011). Firm entry and institutional lock-in: An organizational ecology analysis of the global fashion design industry. *Industrial and Corporate Change*, 20(4), 1031–1048.
Whitaker, J. (2006). *Service and Style: How the American Department Store Fashioned the Middle Class*. New York: St. Martin's Press.
Wiedmann, K. P., Hennigs, N., & Langner, S. (2010). Spreading the word of fashion: Identifying social influencers in fashion marketing. *Journal of Global Fashion Marketing*, 1(3), 142–153.
Wigley, S. M., Moore, C. M., & Birtwistle, G. (2005). Product and brand: Critical success factors in the internationalisation of a fashion retailer. *International Journal of Retail & Distribution Management*, 33(7), 531–544.

Williams, R. & McIlvride, D. (2017). *RiverBlue*. United States: Paddle Productions Inc.
World Bank. (2019). *World Development Indicators. Data Catalogue*; available at: https://datacatalog.worldbank.org/dataset/world-development-indicators
Wyroll, C. (2014). *Social Media: Fundamentals, Models, and Ranking of User–Generated Content*. Berlin: Springer.
Yadav, A., Prasad, V., Kathe, A. A., Raj, S., Yadav, D., Sundaramoorthy, C., & Vigneshwaran, N. (2006). Functional finishing in cotton fabrics using zinc oxide nanoparticles. *Bulletin of materials Science*, 29(6), 641–645.
Yan, M., Ma, X., Yang, Y., Wang, X., Cheong, W. C., Chen, Z., … & Li, Y. (2018). Biofabrication strategy for functional fabrics. *Nano Letters*, 18(9), 6017–6021.
Yang, D. T., Chen, V. W., & Monarch, R. (2010). Rising wages: Has China lost its global labor advantage?. *Pacific Economic Review*, 15(4), 482–504.
Yang, F. C., Wu, K. H., Huang, J. W., Horng, D. N., Liang, C. F., & Hu, M. K. (2012). Preparation and characterization of functional fabrics from bamboo charcoal/silver and titanium dioxide/silver composite powders and evaluation of their antibacterial efficacy. *Materials Science and Engineering: C*, 32(5), 1062–1067.
Yaqub, M., & Lee, W. (2019). Zero-liquid discharge (ZLD) technology for resource recovery from wastewater: A review. *Science of the Total Environment*, 681, 551–563.
Yoon, N., Lee, H. K., & Choo, H. J. (2020). Fast fashion avoidance beliefs and anti-consumption behaviors: The cases of Korea and Spain. *Sustainability*, 12(17), 6907–6927.
Young, S. (28 April, 2018). H&M launches conscious exclusive 2018 collection. *The Independent*. available at: https://www.independent.co.uk/life-style/fashion/hm-conscious-exclusive-2018-collection-sustainable-organic-a8322011.html.
Zafar, A. (2007). The growing relationship between China and Sub-Saharan Africa: Macroeconomic, trade, investment, and aid links. *The World Bank Research Observer*, 22(1), 103–130.
Zarroli, J. (11 March, 2013). *In the trendy world of fast fashion, styles aren't made to last, NPR*, available at: http://www.npr.org/2013/03/11/174013774/in-trendy-world-of-fast-fashion-styles-arent-made-to-last.
Zhang, D. (1997). Inward and outward foreign direct investment: The case of US forest industry. *Forest Products Journal*, 47(5), 29–35.
Zhang, S., Chen, C., Duan, C., Hu, H., Li, H., Li, J., Liu, Y., Ma, X., Stavik, J., & Ni, Y. (2018). Regenerated cellulose by the lyocell process, a brief review of the process and properties. *Bio Resources*, 13(2), 4577–4592.
Zhou, G., Sun, X., & Wang, Y. (2004). Multi-chain digital element analysis in textile mechanics. *Composites Science and Technology*, 64(2), 239–244.
Zhu, S., & Pickles, J. (2014). Bring in, go up, go west, go out: Upgrading, regionalisation and delocalisation in China's apparel production networks. *Journal of Contemporary Asia*, 44(1), 36–63.
Zhu, S., Pickles, J., & He, C. (2017). Turkishization of a Chinese apparel firm: Fast fashion, regionalization, and the shift from global supplier to new end

markets. In *Geographical Dynamics and Firm Spatial Strategy in China* (pp. 97–118). Berlin, Heidelberg: Springer.

Zou, Y., Reddy, N., & Yang, Y. (2011). Reusing polyester/cotton blend fabrics for composites. *Composites Part B: Engineering*, 42(4), 763–770.

Zulfiqar, F., & Thapa, G. B. (2016). Is 'Better cotton'better than conventional cotton in terms of input use efficiency and financial performance?. *Land Use Policy*, 52, 136–143.

INDEX

V

W

Z

Printed in the United States
by Baker & Taylor Publisher Services